# PERGAMON INTERNATIONAL LIBRARY
## of Science, Technology, Engineering and Social Studies
*The 1000-volume original paperback library in aid of education,*
*industrial training and the enjoyment of leisure*
Publisher: Robert Maxwell, M.C.

# Finite Strip Method in
# Structural Analysis

# STRUCTURES AND SOLID BODY MECHANICS SERIES
GENERAL EDITOR: B. G. NEAL

*Other Titles of Interest in the*
**Pergamon International Library**

**AKROYD**
Concrete: Its Properties and Manufacture

**ALLEN**
Analysis and Design of Structural Sandwich Panels

**BROWNE**
Basic Theory of Structures

**CHARLTON**
Model Analysis of Plane Structures

**DUGDALE**
Elements of Elasticity

**DYM**
Introduction to the Theory of Shells

**HENDRY**
Elements of Experimental Stress Analysis

**HEYMAN**
Beams and Framed Structures, 2nd edition

**HORNE & MERCHANT**
The Stability of Frames

**JAEGER**
Elementary Theory of Elastic Plates

**LENCZNER**
Elements of Loadbearing Brickwork
Movement in Buildings

**PARKES**
Braced Frameworks, 2nd edition

The terms of our inspection copy service apply to all the above books. A complete catalogue of all books in the Pergamon International Library is available on request.

# Finite Strip Method in Structural Analysis

by

## Y. K. CHEUNG, D.SC, PH.D, FICE, F.I.STRUCT.E, FIE.AUST.

*Professor of Civil Engineering, and Chairman of Department, University of Adelaide.*
*Formerly Professor of Civil Engineering, University of Calgary*

**PERGAMON PRESS**

OXFORD · NEW YORK · TORONTO
SYDNEY · PARIS · FRANKFURT

| U. K. | Pergamon Press Ltd., Headington Hill Hall, Oxford OX3 0BW, England |
| --- | --- |
| U. S. A. | Pergamon Press Inc., Maxwell House, Fairview Park, Elmsford, New York 10523, U.S.A. |
| C A N A D A | Pergamon of Canada Ltd., P.O. Box 9600, Don Mills M3C 2T9, Ontario, Canada |
| A U S T R A L I A | Pergamon Press (Aust.) Pty. Ltd., 19a Boundary Street, Rushcutters Bay, N.S.W. 2011, Australia |
| F R A N C E | Pergamon Press SARL, 24 rue des Ecoles, 75240 Paris, Cedex 05, France |
| WEST GERMANY | Pergamon Press GmbH, 6242 Kronberg/Taunus, Pferdstrasse 1, Frankfurt-am-Main, West Germany |

First edition 1976

Library of Congress Cataloging in Publication Data

Cheung, Y K
Finite strip method in structural analysis.

(Structures and solid body mechanics series) (Pergamon international library)
Includes bibliographies and index.
1. Structures, Theory of. I. Title. TA646.C48 1975     624'.171     75-19483
ISBN 0-08-018308-5

*Printed in Hungary*

# *Contents*

# *Preface*

IN THIS book I have attempted to present to the reader a concise introduction to the theory of the finite strip method and also the application of the method to the various topics in structural engineering with special reference to practical structures such as slab bridges and box girder bridges. The material included is largely self-contained and the reader is only required to have some basic knowledge of matrix algebra and energy theorems.

The theory of the finite strip method is considered in Chapter 1. From Chapters 2 to 6 many different types of finite strips for plate and shell problems are developed, and numerical examples are given at the end of each chapter. In Chapter 7 the extension of the finite strip method to three-dimensional problems are explored, and in Chapter 8 some computer methods which are commonly used in structural analysis are described. A folded plate computer program is included for completeness, and detailed description for a worked problem is also presented for the sake of clarity.

The finite strip method was developed not very long ago, and it is therefore impossible to claim that all up-to-date materials have been included in this book. A number of publications on the finite strip method have already appeared after the completion of the manuscript.

I am grateful to Professor J. D. Davies, of the University of Swansea, for his encouragement and enthusiasm in my work while I was Lecturer in Swansea, and for his suggestion that this book should be written. I would also like to express my gratitude to Professor Y. Niwa of Kyoto University, and Professor J. J. Raftery of the University of Hong Kong for providing me with facilities to write the book while I was on sabbatical leave in the above two institutions. I am obliged to Dr. D. Ho of the University of Hong Kong for her comments and criticisms on several chapters of the manuscripts. Finally, I am indebted to Miss S. L. Chan for preparation of the major part of the typescript.

# Nomenclature

| | |
|---|---|
| $a, b,$ | strip length and width |
| $A_i$ | undetermined constants of the polynomial displacement functions |
| $[B]$ | strain matrix |
| $c$ | thickness of a cylindrical layer |
| $C_i$ | undetermined constants of the basic functions |
| $[C_i]$ | shape functions for a strip |
| $D_x, D_y, D_{xy}$ $D_r, D_\theta, D_k$ | bending and twisting rigidities in Cartesian and polar coordinates |
| $[D]$ | property matrix |
| $E_x, E_y, E_{xy}$ $E_r, E_\theta, E_{r\theta}$ | elastic constants in Cartesian and polar coordinates |
| $e$ | eccentricity |
| $f(x), f(r)$ | polynomial functions in transverse (Cartesian) and radial (polar) directions |
| $f_i$ | nodal parameters of a strip |
| $\{F\}$ | load vector |
| $G, G_{xy}, G_{r\theta}$ | shear moduli |
| $h_j$ | thickness of sandwich core |
| $[H]$ | transformation matrix |
| $I$ | moment of inertia |
| $I_i$ | integrals |
| $[J]$ | Jacobian matrix |
| $[K]$ | assembled stiffness matrix |
| $[K_G]$ | assembled geometric stiffness matrix |
| $k_{mn}^f, k_{mn}^t$ | flexural and torsional stiffness coefficients of a beam element |
| $L_i$ | area coordinate |
| $[M]$ | mass matrix |

| | |
|---|---|
| $M_x, M_y, M_{xy}$ $M_r, M_\theta, M_{r\theta}$ | bending and twisting moments in Cartesian and polar coordinates |
| $[N_k]$ | shape functions |
| $P$ | concentrated load |
| $\{P\}$ | load vector |
| $q$ | uniformly distributed load |
| $\{q\}$ | external surface load vector |
| $R, r$ | radius of the curve strip plate |
| $[R]$ | transformation matrix |
| $[S]$ | stiffness matrix of strip |
| $[S_G]$ | geometric stiffness matrix of strip |
| $t$ | thickness of plates |
| $t_i$ | thickness of $i$th layer of sandwich plates |
| $u, v, w$ | displacements in $x$, $y$, and $z$ directions respectively |
| $U$ | strain energy |
| $W$ | potential energy due to external forces |
| $x, y, z$ $r, \theta$ | rectangular and polar coordinates |
| $X_m, Y_m$ | basic functions |
| $\{X\}$ | redundant force vector |
| $\alpha$ | subtend angle |
| $\beta$ | angle of skew |
| $\varepsilon, \varepsilon_x, \varepsilon_y, \gamma_{xy}$ | strains |
| $\sigma, \sigma_x, \sigma_y, \tau_{xy}$ | stress |
| $\{\sigma\}, \{\varepsilon\}, \{\chi\}$ | generalized stress and strain vectors |
| $\chi_i$ | curvature parameters |
| $\theta_x, \theta_y$ | rotations about $x$ and $y$ directions respectively |
| $\mu_m, \alpha_m$ | parameters of the basic functions |
| $\delta_k$ | nodal displacement parameters |
| $\{\delta\}$ | nodal displacement vector |
| $\xi, \eta, \zeta$ | skew or other curvilinear coordinate systems |
| $\phi$ | total potential energy |
| $\nu$ | poisson's ratio |
| $\omega$ | natural frequency |
| $\varrho$ | mass density |
| $\lambda$ | eigenvalues |

# CHAPTER 1

# *Finite strip method*

## 1.1. INTRODUCTION

The finite element method,[1] as the most powerful and versatile tool of
solution in structural analysis, is now well known and established. However, for many structures having regular geometric plans and simple
boundary conditions, a full finite element analysis is very often both
extravagant and unnecessary, and at times even impossible. The cost of
solutions can be very high, and usually jumps by an order of magnitude
when a more refined, higher dimensional analysis is required. Moreover,
very often the problem size of an accurate analysis might be so over-
whelming as to overtax whatever machines that are available to a designer or researcher, so that the problem either will have to be solved
roughly, or some additional lengthy, time-consuming subroutines written
to lower the core requirements. The above observations are especially
true for static analysis of three-dimensional solids and spatial structures
and for eigenvalue problems in vibration and stability analysis. An alternative method which can reduce the computational effort and core
requirements, but at the same time retaining to some extent the versatility
of the finite element analysis, is evidently desirable for the afore-mentioned
class of structures.

These requirements can be satisfied fully by the recently developed
finite strip method. In this method, the structure is divided into two
(strips) or three-dimensional (prisms, layers) subdomains in which one
opposite pair of the sides (2-D) or one or more opposite pairs of faces
(3-D) of such a subdomain are in coincidence with the boundaries of the
structure. The geometry of the structure is usually constant along one
or two coordinate axes so that the width of a strip or the cross-section
of a prism or layer will not change from one end to the other. Thus

while box girder bridges and voided slabs are conveniently divided into strips or prisms, for thick, isotropic, or multi-layered plates and shells a division into layers would definitely be more suitable.

The finite strip method can be considered as a special form of the finite element procedure using the displacement approach. Unlike the standard finite element method, which uses polynomial displacement functions in all directions, the finite strip method calls for use of simple polynomials in some directions and continuously differentiable smooth series in the other directions, with the stipulation that such series should satisfy *a priori* the boundary conditions at the ends of the strips or prisms. The general form of the displacement function is given as a product of polynomials and series. Thus for a strip, in which a two-dimensional problem is reduced to a one-dimensional problem,

$$w = \sum_{m=1}^{r} f_m(x)\, Y_m. \tag{1.1a}$$

Similarly, for the case of a prism, a three-dimensional problem is reduced to that of a two-dimensional one, and the displacement function is written as

$$w = \sum_{m=1}^{r} f_m(x, z)\, Y_m. \tag{1.1b}$$

Finally, a three-dimensional problem is treated as a one-dimensional one in the case of a layer, and

$$w = \sum_{m=1}^{r} \sum_{n=1}^{t} f_{mn}(z)\, X_m Y_n. \tag{1.1c}$$

In the above expressions, the series has been truncated at $r$th and $t$th terms; $f_m(x)$, $f_m(x, z)$, $f_{mn}(z)$ are polynomial expressions with undetermined constants for the $m$th and $n$th terms of the series; and $X_m$, $Y_n$ are series which satisfy the end conditions in the $x$ and $y$ directions respectively and also specify the deflected shapes in those directions. A diagrammatic explanation of the above discussions and the effect of reduced dimension on matrix bandwidth is found in Fig. 1.1 for some practical structures. A brief comparison between the finite element method and the finite strip method is listed in Table 1.1.

TABLE 1.1. COMPARISON BETWEEN FINITE ELEMENT AND FINITE STRIP METHODS

| Finite element | Finite strip |
|---|---|
| Applicable to any geometry, boundary conditions and material variation. Extremely versatile and powerful | In static analysis, more often used for structures with two opposite simply supported ends and with or without intermediate elastic supports, especially for bridges. In dynamic analysis it is used for structures with all boundary conditions but without discrete supports |
| Usually large number of equations and matrix with comparatively large bandwith. Can be very expensive and at times impossible to work out solution because of limitation in computing facilities | Usually much smaller number of equations and matrix with narrow bandwidth, especially true for problems with an opposite pair of simply supported ends. Consequently much shorter computing time for solution of comparable accuracy |
| Large quantities of input data and easier to make mistakes. Requires automatic mesh and load generation schemes | Very small amount of input data because of the small number of mesh lines involved due to the reduction in dimensional analysis |
| Large quantities of output because as a rule all nodal displacements and element stresses are printed. Also many lower order elements will not yield correct stresses at the nodes and stress averaging or interpolation techniques must be used in the interpretation of results | Easy to specify only those locations at which displacements and stresses are required and then output accordingly |
| Requires a large amount of core and is more difficult to program. Very often, advanced techniques such as mass condensation or subspace iteration have to be resorted to for eigenvalue problems in order to reduce core requirements | Requires smaller amount of core and is easier to program. Because only the lowest few eigenvalues are required (for most cases anyway), the first two to three terms of the series will normally yield sufficiently accurate results. Matrix can usually be solved by standard eigenvalue subroutines |

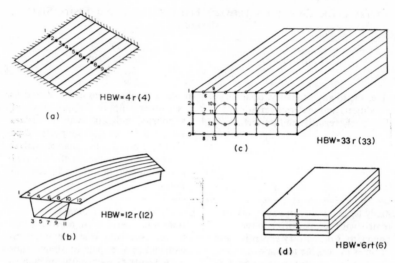

FIG. 1.1. Some structures and typical mesh divisions. (a) Encastred slab
(plate strips). (b) Curved box girder bridge (shell strips). (c) Voided slab
bridge with circular holes (quadrilateral finite prisms). (d) Thick, multi-
layer plate (finite layers). HBW = half bandwidth; $r$, $t$, = number of
terms of series; ( ) = simply supported case.

The philosophy of the finite strip method is similar to that of the Kantorovich[2] method which is used extensively for reducing a partial differential equation to an ordinary differential equation.

The finite strip method requires the discretization of the continuum
so that only a finite number of unknowns will exist in the resulting formulation. The following procedure is adopted:

(i) The continuum is divided into strips (prisms, layers) by fictitious
lines (surfaces). The ends of such strips (prisms, layers) always constitute
a part of the boundaries of the continuum (Fig. 1.1). For convenience,
only the strip will be discussed in the general formulation, which is in
any case also applicable to prisms and layers.

(ii) The strips are assumed to be connected to one another along a
discrete number of nodal lines which coincide with the longitudinal boundaries of the strip. In some cases it is also possible to use internal nodal
lines to arrive at a higher order strip.

The degrees of freedom (DOF) at each nodal line, called nodal displacement parameters, are normally connected with the displacements and their first partial derivatives (rotations) with respect to the polynomial variable $x$ in the transverse direction. They can also include non-displacement terms such as strains (including direct strains, shear strain, bending, and twisting curvatures).

Due to the use of continuous functions in the longitudinal direction, the DOF at a strip nodal line is usually less than that at an element node. For example, in plate bending, $w$ and $\theta_x$ exist at each strip nodal line, while $w$, $\theta_x$, $\theta_y$ exist at each element node.

(iii) A displacement function (or functions), in terms of the nodal displacement parameters, is chosen to represent the displacement field and consequently the strain and stress (including direct stresses, shear stress, bending, and twisting moments) fields within each element.

(iv) Based on the chosen displacement function, it is possible to obtain a stiffness matrix and load matrices which equilibrate the various concentrated or distributed loads acting on the strip through either virtual work or minimum total potential energy principles.

(v) The stiffness and load matrices of all the strips are assembled to form a set of overall stiffness equations. Since the bandwidth and the size of the matrix is usually small, the equations can be solved easily by any standard band matrix solution technique to yield the nodal displacement parameters. In fact, once the stiffness and load matrices have been developed for a strip, it can be said that the finite strip analysis of a plate problem will be the same as the stiffness analysis of a beam problem, while that of a shell or box girder bridge analysis will correspond to a plane frame analysis.

## 1.2. CHOICE OF DISPLACEMENT FUNCTIONS

From the previous discussion it is seen that the choice of suitable displacement functions for a strip is the most crucial part of the analysis, and great care must be exercised at such a stage. A wrongly chosen displacement function might not just produce obviously ridiculous answers but may even lead to results which converge to the wrong answer for successively refined meshes.[3]

To ensure convergence to the correct results the following simple requirements have to be satisfied.

(i) The series part ($Y_m$) of the displacement function should satisfy *a priori* the end conditions of the strip (for vibration problems only the displacement conditions have to be satisfied). For example, for a simply supported plate strip in bending, the displacement function should be able to satisfy the conditions of both deflection $w$ and normal curvature $\partial^2 w/\partial y^2$ being equal to zero at the two ends.

(ii) The polynomial part of the displacement function ($f_m(x)$) must be able to represent a state of constant strain in the transverse ($x$) direction. If this is not obeyed, then there is no guarantee that the strain will converge towards the true strain distribution as the mesh is further and further subdivided.

This constant strain condition can be tested in one of the following two ways:

(a) If a simple polynomial of the form $a_1 + a_2 x + a_3 x^2 + \ldots$ is used as $f(x)$, then constant strain will exist if the polynomial is complete up to or above the order in which a constant term will actually be obtained when the necessary differentiation for computing strains are carried out. For example, in the case of a bending strip, the polynomial must be complete up to the quadratic term at least. If the polynomial is chosen as $a_1 + a_2 x + a_3 x^3 + a_4 x^4$, then the transverse bending curvature $\chi_x = -\partial^2 w/\partial x^2$ will be equal to $-6a_3 x - 12a_4 x^2$. This would naturally imply that $\chi_x$ is always zero at $x = 0$, and that $\chi_x$ will never converge towards the correct solution.

(b) If shape functions (see Section 1.3.2.) are used, it is not possible to decide very easily whether the polynomial is complete or not, and a different approach has to be resorted to.

In general the nodal displacement parameters of a deformed strip will take up arbitrary values which bear no fixed relationship to each other. However, for a constant strain situation such as cylindrical bending, pure shear, etc., the nodal displacement parameters will be related to each other in some specific manner on taking up some prescribed values. If such a set of parameters compatible with a constraint condition is substituted into the displacement function, then constant strain should in fact be obtained if the shape functions are correctly formulated and well behaved.

For example, the displacement function for a bending plate strip is given by

$$\{f\} = \{w\} = \sum_{m=1}^{r} [(1-3\bar{x}^2+2\bar{x}^3) w_{1m}+(x-2\bar{x}x+\bar{x}^2x) \theta_{1m}$$
$$+(3\bar{x}^2-2\bar{x}^3) w_{2m}+(\bar{x}^2x-\bar{x}x) \theta_{2m}] Y_m \qquad (1.2)$$

in which $\bar{x} = x/b$ and $b$ is the width of strip.

For $w_{1m} = w_{2m} = 0$ and $\theta_{1m} = -\theta_{2m}$, a state of constant curvature in the $x$ direction can be expected. If these nodal displacement parameters are substituted into (1.1), then

$$w = \sum_{m=1}^{r} (x-\bar{x}x) \theta_{1m}Y_m \qquad (1.3a)$$

and

$$\frac{\partial^2 w}{\partial x^2} = \sum_{m=1}^{r} \frac{-2}{b} \theta_{1m}Y_m. \qquad (1.3b)$$

Thus a state of constant bending strain in the $x$ direction in fact exists.

(iii) The displacement function must satisfy the compatibility of displacements along the boundaries with neighbouring strips, and this might include the continuity of the first partial derivative as well as the displacement values.

The above statement can also be rephrased as "the displacement function should be chosen in such a way that the strains which are required in the energy formulations should remain finite at the interface between the strips".

Thus in two-dimensional elasticity the strains involved are first partial derivatives, and therefore only the displacements need to be continuous. On the other hand, for bending problems the strains involved are second partial derivatives, and both the displacements and their first partial derivatives will have to be continuous at the interface.

If such conditions are complied with, then there will be no infinite strains at the interface and therefore no contribution to the energy formulation from the interface, which can be considered as a narrow strip of area converging to zero. Only in this way can we be assured that a simple summation of the total potential energy of all the strips will in fact be equal to the total potential energy of the elastic body in question. The total potential energy of such a finite strip representation will always provide an approximate energy greater than the true one, therefore giving a bound to the absolute total potential energy of the elastic system. A detailed and mathematical discussion of this condition was presented in a paper by Tong and Pian.[4]

## 1.3. AVAILABLE DISPLACEMENT FUNCTIONS

Since a displacement function is always made up of two parts [see (1.1a)], a polynomial $f_m(x)$ governed by the shape of the cross-section (e.g. line, triangle, etc.), together with the nodal arrangement within the cross-section, and a series $Y_m$ determined by the end conditions, it would be convenient for us to discuss each part separately.

### 1.3.1. SERIES PART OF DISPLACEMENT FUNCTION

The most commonly used series are the basic functions (or eigenfunctions) which are derived from the solution of the beam vibration differential equation

$$Y''''  = \frac{\mu^4}{a^4} Y, \tag{1.4}$$

where $a$ is length of beam (strip) and $\mu$ is a parameter.

The general form of the basic functions is

$$Y(y) = C_1 \sin\left(\frac{\mu y}{a}\right) + C_2 \cos\left(\frac{\mu y}{a}\right) + C_3 \sinh\left(\frac{\mu y}{a}\right) + C_4 \cosh\left(\frac{\mu y}{a}\right) \tag{1.5}$$

with the coefficients $C_1$, etc., to be determined by the end conditions. These have been worked out explicitly in the literature [5] for the various end conditions and are listed below:

(a) Both ends simply supported $(Y(0) = Y''(0) = 0, Y(a) = Y''(a) = 0)$.[†]

$$Y_m(y) = \sin\left(\frac{\mu_m y}{a}\right) \quad (\mu_m = \pi, 2\pi\ 3\pi, \ldots m\pi). \tag{1.6a}$$

---

[†] The prime and double prime above $Y$ refer to the first and second derivative of $Y$ respectively.

(b) Both ends clamped $(Y'(0) = Y'(0) = 0, \; Y(a) = Y'(a) = 0)$.

$$\left.\begin{aligned}
Y_m(y) &= \sin\left(\frac{\mu_m y}{a}\right) - \sinh\left(\frac{\mu_m y}{a}\right) - \alpha_m\left[\cos\left(\frac{\mu_m y}{a}\right) - \cosh\left(\frac{\mu_m y}{a}\right)\right], \\[2mm]
\alpha_m &= \frac{\sin\mu_m - \sinh\mu_m}{\cos\mu_m - \cosh\mu_m} \\[2mm]
&\left(\mu_m = 4.7300,\, 7.8532,\, 10.9960,\, \ldots\, \frac{2m+1}{2}\,\pi\right).
\end{aligned}\right\}$$

(1.6b)

(c) One end simply supported and the other end clamped

$$(Y(0) = Y''(0) = 0, \; Y(a) = Y'(a) = 0).$$

$$\left.\begin{aligned}
Y_m(y) &= \sin\left(\frac{\mu_m y}{a}\right) - \alpha_m \sinh\left(\frac{\mu_m y}{a}\right), \\[2mm]
\alpha_m &= \frac{\sin\mu_m}{\sinh\mu_m} \\[2mm]
&\left(\mu_m = 3.9266,\, 7.0685,\, 10.2102,\, \ldots,\, \frac{4m+1}{4}\,\pi\right)
\end{aligned}\right\}$$

(1.6c)

(d) Both ends free $(Y''(0) = Y'''(0) = 0, \; Y''(a) = Y'''(a) = 0)$.

$$\left.\begin{aligned}
Y_1(y) &= 1, \quad \mu_1 = 0, \\
Y_2(y) &= 1 - \frac{2y}{a}, \quad \mu_2 = 1, \\[1mm]
Y_m(y) &= \sin\frac{\mu_m y}{a} + \sinh\frac{\mu_m y}{a} - \alpha_m\left(\cos\frac{\mu_m y}{a} + \cosh\frac{\mu_m y}{a}\right), \\[2mm]
\alpha_m &= \frac{\sin\mu_m - \sinh\mu_m}{\cos\mu_m - \cosh\mu_m} \\[2mm]
&\left(\mu_m = 4.7300,\, 7.8532,\, 10.9960,\, \ldots,\, \frac{2m-3}{2}\,\pi,\; m = 3,\; 4,\, \ldots,\, \infty\right).
\end{aligned}\right\}$$

(1.6d)

(e) One end clamped and the other end free $(Y(0) = Y'(0) = 0,$
$Y''(a) = Y'''(a) = 0)$.

$$\left.\begin{aligned}
Y_m(y) &= \sin \frac{\mu_m y}{a} + \sinh \frac{\mu_m y}{a} - \alpha_m \left(\cos \frac{\mu_m y}{a} - \cosh \frac{\mu_m y}{a}\right), \\
\alpha_m &= \frac{\sin \mu_m + \sinh \mu_m}{\cos \mu_m + \cosh \mu_m} \\
&\left(\mu_m = 1.875, 4.694, \ldots, \frac{2m-1}{2} \pi\right).
\end{aligned}\right\} \quad (1.6e)$$

(f) One end simply supported and the other end free
$$(Y(0) = Y''(0) = 0, \quad Y''(a) = Y'''(a) = 0).$$

$$\left.\begin{aligned}
Y_1(y) &= \frac{y}{a}, \quad \mu_1 = 1, \\
Y_m(y) &= \sin \frac{\mu_m y}{a} + \alpha_m \sinh \frac{\mu_m y}{a}, \\
\alpha_m &= \frac{\sin \mu_m}{\sinh \mu_m} \\
&\left(\mu_m = 3.9266, 7.0685, 10.2102, 13.3520, \ldots, \frac{2m-3}{4} \pi, \right. \\
&\left. \qquad m = 2, 3, \ldots, \infty\right).
\end{aligned}\right\} \quad (1.6f)$$

The above functions are primarily used for bending strips. For two- and three-dimensional elasticity problems, which will be discussed in more detail in Chapter 3, both $Y_m$ and $Y'_m$ will be used for the $u$, $v$ (and $w$) displacements. Only one other function has been employed successfully in plane analysis and will be discussed later.

The basic functions possess the valuable properties of orthogonality, i.e.

$$\left.\begin{aligned}
\int_0^a Y_m Y_n \, dy &= 0 \\
\int_0^a Y''_m Y''_n \, dy &= 0
\end{aligned}\right\} \text{ for } m \neq n. \quad (1.7)$$

It will be observed that these integrals appear in all subsequent formulations, and the utilization of these properties results in a significant saving in computation effort.

The existence of so many functions should in no way be interpreted as a need for a matching number of computer subroutines. In actual practice all the relevant functions are stored in the same subroutine, and integrals involving different combinations of the functions are integrated numerically. Consequently problems with many different types of boundary conditions can be solved by the same program.

## 1.3.2. SHAPE FUNCTION PART OF DISPLACEMENT FUNCTION

A shape function is a polynomial associated with a nodal displacement parameter, and it describes the corresponding displacement field within the cross-section of a strip when the nodal displacement parameter in question is given unit value. In fact such shape functions are derived by specifying a normalized unit value of the relevant displacement component at its own node, and a value of zero for the same displacement component at all other nodes.

For example, (1.2) can be written as

$$w = \sum_{m=1}^{r} [[C_1] [C_2]] \begin{Bmatrix} \{\delta_1\} \\ \{\delta_2\} \end{Bmatrix}_m Y_m$$

$$= \sum_{m=1}^{r} Y_m \sum_{k=1}^{2} [C_k] \{\delta_k\}_m \qquad (1.8)$$

in which $\begin{Bmatrix} \{\delta_1\} \\ \{\delta_2\} \end{Bmatrix}_m$ is a vector representing the $m$th term nodal displacement parameters (deflection and rotation) at nodes 1 and 2, and $[C_1]$, $[C_2]$ are the shape functions associated with $\{\delta_1\}$ and $\{\delta_2\}$ respectively.

By virtue of (1.2) and (1.8), we find that at $x = 0$

$$[[C_1] [C_2]] = [1 \ 0 \ 0 \ 0],$$

$$\left[ \frac{\partial [C_1]}{\partial x} \ \frac{\partial [C_2]}{\partial x} \right] = [0 \ 1 \ 0 \ 0],$$

and at $x = b$

$$[[C_1] \, [C_2]] = [0 \ 0 \ 1 \ 0],$$

$$\left[ \frac{\partial[C_1]}{\partial x} \ \frac{\partial[C_2]}{\partial x} \right] = [0 \ 0 \ 0 \ 1],$$

thus satisfying the stated criterion.

The main purpose of using shape functions directly instead of a simple polynomial with undetermined constants is twofold: to avoid the lengthy process of relating the undetermined constants to the nodal displacement parameters, and to make sure that there is compatibility of displacements along the interface of adjoining strips (prisms). The first point is rather obvious and requires no discussion. The second point can best be explained by noting that the displacements along any interface of adjoining strips (prisms) are uniquely determined by the displacement parameters at the node (or nodes) common to the adjoining strips (prisms), since by definition the shape function for the displacement parameters of any other node will take up zero values at the said interface. Many shape functions are available, and some of the more common ones are listed below.

(a) Straight line with two nodes (Fig. 1.2a) and with displacements as nodal parameters:

$$C_1 = (1-\bar{x}), \quad C_2 = \bar{x}. \tag{1.9a}$$

(b) Straight line with two nodes (Fig. 1.2b); displacements and first derivatives:

$$\left. \begin{array}{l} [C_1] = [(1-3\bar{x}^2+2\bar{x}^3), \quad x(1-2\bar{x}+\bar{x}^2)], \\ [C_2] = [(3\bar{x}^2-2\bar{x}^3), \quad\quad x(\bar{x}^2-\bar{x})]. \end{array} \right\} \tag{1.9b}$$

(c) Straight line with two nodes (Fig. 1.2c); displacements, first and second derivatives:

$$[C_1] = [(1-10\bar{x}^3+15\bar{x}^4-6\bar{x}^5), \quad x(1-6\bar{x}^2+8\bar{x}^3-3\bar{x}^4), \quad x^2(0.5-1.5\bar{x}$$
$$+ 1.5\bar{x}^2-0.5\bar{x}^3)],$$

$$\tag{1.9c}$$

$$[C_2] = [(10\bar{x}^3-15\bar{x}^4+6\bar{x}^5), \quad x(-4\bar{x}^2+7\bar{x}^3-3\bar{x}^4), \quad x^2(0.5\bar{x}-\bar{x}^2+0.5\bar{x}^3)]$$

(d) Straight line with three nodes (Fig. 1.2d); displacements only:

$$C_1 = (1-3\bar{x}+2\bar{x}^2), \quad C_2 = (4\bar{x}-4\bar{x}^2), \quad C_3 = (-\bar{x}+2\bar{x}^2). \tag{1.9d}$$

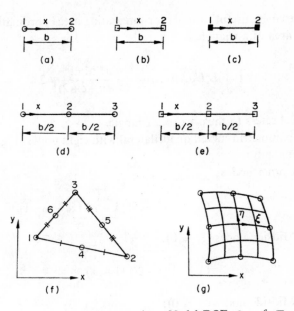

FIG. 1.2. Strip and prism cross-sections. Nodal DOF: $\bigcirc = f$; $\square = f$,
$$\frac{\partial f}{\partial x} \; ; \; \blacksquare = f, \frac{\partial f}{\partial x}, \frac{\partial^2 f}{\partial x^2} \; .$$

(e) Straight line with three nodes (Fig. 1.2e); displacements and first derivatives:

$$\left.\begin{aligned}
[C_1] &= [(1-23\bar{x}^2+66\bar{x}^3-68\bar{x}^4+24\bar{x}^5), \quad x(1-6\bar{x}+13\bar{x}^2-12\bar{x}^3+4\bar{x}^4)], \\
[C_2] &= [(16\bar{x}^2-32\bar{x}^3+16\bar{x}^4), \quad x(-8\bar{x}+32\bar{x}^2-40\bar{x}^3+16\bar{x}^4)], \\
[C_3] &= [(7\bar{x}^2-34\bar{x}^3+52\bar{x}^4-24\bar{x}^5), \quad x(-\bar{x}+5\bar{x}^2-8\bar{x}^3+4\bar{x}^4)].
\end{aligned}\right\}$$
(1.9e)

(f) Triangle with six nodes (Fig. 1.2f); displacements only:
(1) Corner nodes:

$$C_i = (2L_i-1)L_i \quad (i = 1, 2, 3).$$

(2) Midside nodes:

$$C_4 = 4L_1L_2, \quad C_5 = 4L_2L_3, \quad C_6 = 4L_3L_1,$$
(1.9f)

in which $L_1$, etc, are area coordinates.[1]

An extremely useful formula for integrating area coordinate quantities over the area of a triangle is

$$\int\int_{\Delta} L_1^a L_2^b L_3^c \, dx \, dy = \frac{a! \, b! \, c!}{(a+b+c+2)!} \, 2\Delta$$

in which $\Delta$ refers to the area of a triangle.

(g) Isoparametric curved quadrilateral with eight nodes (Fig. 1.2g); displacements only:

(1) Corner nodes:

$$C_i = \tfrac{1}{4}(1+\xi_0)(1+\eta_0)(\xi_0+\eta_0-1).$$

(2) Midside nodes ($\xi_i = 0$):

$$C_i = \tfrac{1}{2}(1-\xi^2)(1+\eta_0).$$

(3) Midside nodes ($\eta_i = 0$):

$$C_i = \tfrac{1}{2}(1+\xi_0)(1-\eta^2) \tag{1.9g}$$

with $\xi_0 = \xi\xi_i$, $\eta_0 = \eta\eta_i$, and $\xi_i$ and $\eta_i$ being the $\xi$ and $\eta$ coordinates of the $i$th node.

A detailed discussion on isoparametric elements was presented by Zienkiewicz et al.[6]

The derivation of shape functions merits some close attention. In general, one of the two methods outlined here can be adopted, and case (d) of Fig. 1.2d (Section 1.3.2) will be used as an example in all subsequent discussions.

(i) *Simple polynomial and direct inversion*

There are altogether 3 DOF for case (d), and hence a second degree polynomial with three undetermined constants would be adequate for describing the displacements.

$$f = A_1 + A_2 x + A_3 x^2 \tag{1.10}$$

or, in matrix form,

$$f = [1 \quad x \quad x^2] \begin{Bmatrix} A_1 \\ A_2 \\ A_3 \end{Bmatrix}. \tag{1.11}$$

Applying (1.11) to the three nodes successively, we obtain

$$\begin{Bmatrix} f_1 \\ f_2 \\ f_3 \end{Bmatrix} = \begin{bmatrix} 1 & x_1 & x_1^2 \\ 1 & x_2 & x_2^2 \\ 1 & x_3 & x_3^2 \end{bmatrix} \begin{Bmatrix} A_1 \\ A_2 \\ A_3 \end{Bmatrix} = \begin{bmatrix} 1 & 0 & 0 \\ 1 & \dfrac{b}{2} & \dfrac{b^2}{4} \\ 1 & b & b^2 \end{bmatrix} \begin{Bmatrix} A_1 \\ A_2 \\ A_3 \end{Bmatrix} \tag{1.12}$$

in which $f_1$, etc, are the values of $f$ at the three nodes.

The constants $\{A\}$ can be expressed in terms of the nodal parameters $f_i$ by the inversion of (1.12). This is done numerically by computer or algebraically as follows:

(a) Write out (1.12) in longhand:

$$f_1 = A_1, \tag{1.13a}$$

$$f_2 = A_1 + A_2\frac{b}{2} + A_3\frac{b^2}{4}, \tag{1.13b}$$

$$f_3 = A_1 + A_2 b + A_3 b^2. \tag{1.13c}$$

(b) Eliminate $A_2$ by performing $[(1.13c) - 2 \times (1.13b)]$; then

$$A_3 = \frac{2}{b^2}(f_1 - 2f_2 + f_3). \tag{1.13d}$$

(c) Obtain $A_2$ by substituting (1.13d) into (1.13c):

$$A_2 = \frac{1}{b}(-3f_1 + 4f_2 - f_3). \tag{1.13e}$$

(d) Put (1.13a), (1.13d), and (1.13e) in matrix form:

$$\begin{Bmatrix} A_1 \\ A_2 \\ A_3 \end{Bmatrix} = \begin{bmatrix} 1 & 0 & 0 \\ \dfrac{-3}{b} & \dfrac{4}{b} & \dfrac{-1}{b} \\ \dfrac{2}{b^2} & \dfrac{-4}{b^2} & \dfrac{2}{b^2} \end{bmatrix} \begin{Bmatrix} f_1 \\ f_2 \\ f_3 \end{Bmatrix}. \tag{1.13f}$$

(e) Equation (1.13f) is now substituted into (1.11) to yield

$$f = \begin{bmatrix} 1 & x & x^2 \end{bmatrix} \begin{bmatrix} 1 & 0 & 0 \\ \dfrac{-3}{b} & \dfrac{4}{b} & \dfrac{-1}{b} \\ \dfrac{2}{b^2} & \dfrac{-4}{b^2} & \dfrac{2}{b^2} \end{bmatrix} \begin{Bmatrix} f_1 \\ f_2 \\ f_3 \end{Bmatrix}. \tag{1.13g}$$

The product of the first two matrices will obviously give the shape functions at all the three nodes. Thus if the matrix multiplication operations are carried out,

$$f = \left[ \left(1 - \frac{3x}{b} + \frac{2x^2}{b^2}\right), \quad \left(\frac{4x}{b} - \frac{4x^2}{b^2}\right), \quad \left(\frac{-x}{b} + \frac{2x^2}{b^2}\right) \right] \begin{Bmatrix} f_1 \\ f_2 \\ f_3 \end{Bmatrix}$$

$$= \left[ (1 - 3\bar{x} + 2\bar{x}^2), \quad (4\bar{x} - 4\bar{x}^2), \quad (-\bar{x} + 2\bar{x}^2) \right] \begin{Bmatrix} f_1 \\ f_2 \\ f_3 \end{Bmatrix}$$

$$= \begin{bmatrix} C_1 & C_2 & C_3 \end{bmatrix} \begin{Bmatrix} f_1 \\ f_2 \\ f_3 \end{Bmatrix}. \tag{1.13h}$$

Thus the shape functions of (1.9d) have been arrived at by assuming a simple polynomial and subsequently going through the process of relating the undetermined constants to the nodal parameters.

The above procedure requires little ingenuity and is found to be convenient and useful in establishing shape functions for simpler cases. For more complex cases, the algebraic inversion process outlined in (1.13) becomes very lengthy and involved and a direct formulation of the shape functions (see below) very often will be more advantageous.[†] The reader should also watch out for the occasional cases in which an inverse simply does not exist.

## (ii) *Direct formulation of shape functions*

In this direct approach, Lagrange and Hermitian polynomials have been used for creating special families of shape functions.[7] For the gen-

---

† The problem of algebraic inversion can be alleviated to a certain extent by the use of FORMAC,[8] which is a computer language for doing algebraic operations.

eral case, however, the shape functions are very often simply obtained by inspection.

Again take case (d) as an example. For compatibility requirements it is necessary to satisfy the conditions of

at node 1, $\qquad C_1 = 1$ (1.14a)

and at nodes 2 and 3 $\qquad C_1 = 0,$ (1.14b)

etc. Linear interpolation functions are used as a basis and are combined in such a way so that the conditions set forth in (1.14) can be realized. The process of constructing suitable shape functions is summarized in

TABLE 1.2. THE CONSTRUCTION OF SHAPE FUNCTIONS FOR CASE (d)

| | Trial functions | Node 1 | Node 2 | Node 3 | Eqn. (1.14) satisfied |
|---|---|---|---|---|---|
| $C_1$ | $\left(1-\dfrac{x}{b}\right)$ | 1 | $\frac{1}{2}$ | 0 | No |
| | $\left(1-\dfrac{2x}{b}\right)$ | 1 | 0 | $-1$ | No |
| | $\left(1-\dfrac{x}{b}\right)\left(1-\dfrac{2x}{b}\right)=(1-\bar{x})(1-2\bar{x})$ | 1 | 0 | 0 | Yes |
| $C_2$ | $2\dfrac{x}{b}$ | 0 | 1 | 2 | No |
| | $2\left(1-\dfrac{x}{b}\right)$ | 2 | 1 | 0 | No |
| | $4\dfrac{x}{b}\left(1-\dfrac{x}{b}\right)=4\bar{x}(1-\bar{x})$ | 0 | 1 | 0 | Yes |
| $C_3$ | $\dfrac{x}{b}$ | 0 | $\frac{1}{2}$ | 1 | No |
| | $-\left(1-\dfrac{2x}{b}\right)$ | $-1$ | 0 | 1 | No |
| | $-\dfrac{x}{b}\left(1-\dfrac{2x}{b}\right)=-\bar{x}(1-2\bar{x})$ | 0 | 0 | 1 | Yes |

Table 1.2. Note that a linear function having zero value at a particular node will cause the combined function to have zero value at the same node.

## 1.4. FORMULATION OF STRIP CHARACTERISTICS THROUGH MINIMUM TOTAL POTENTIAL ENERGY PRINCIPLE

The formulation of strip characteristics will now be presented in detail using the principle of minimum total energy which is well known in structural mechanics. The principle states that "of all compatible displacements satisfying given boundary conditions, those which satisfy the equilibrium conditions make the total potential energy assume a stationary value". In mathematical form we have

$$\left\{\frac{\partial\phi}{\partial\{\delta\}}\right\} = \left\{\begin{array}{c} \dfrac{\partial\phi}{\partial\{\delta\}_1} \\[2mm] \dfrac{\partial\phi}{\partial\{\delta\}_2} \\[1mm] \vdots \end{array}\right\} = \{0\} \qquad (1.15)$$

in which the total potential energy $\phi$ is defined as the sum of the potential energy of external forces $W$ and the strain energy $U$. $\{\delta\}_m$ is a vector of nodal displacement parameters at all nodes for the $m$th term of the series.

Equation (1.15) can also be interpreted as the more general Raleigh–Ritz approach applied to elastic analysis.

### 1.4.1. DISPLACEMENT FUNCTIONS

The general form of displacement functions can be written as

$$\{f\} = \sum_{m=1}^{r} [[C_1]\,[C_2]\,\ldots] \left\{\begin{array}{c} \{\delta_1\} \\ \{\delta_2\} \\ \vdots \end{array}\right\}_m Y_m$$

$$= \sum_{m=1}^{r} Y_m \sum_{k=1}^{s} [C_k]\,\{\delta_k\}_m. \qquad (1.16a)$$

The expressions can be made more compact by combining together the series and the shape functions. In this way

$$\{f\} = \sum_{m=1}^{r} \sum_{k=1}^{s} [N_k]_m \{\delta_k\}_m$$

$$= \sum_{m=1}^{r} [N]_m \{\delta\}_m \qquad (1.16b)$$

$$= [N] \{\delta\}$$

in which $k = 1, 2, \ldots, s$ refer to the nodal line numbers.

For the case of a plane stress strip with two nodal lines and $u, v$ as nodal displacement parameters, the following equalities would apply:

$$\left.\begin{array}{c} \{f\} = \begin{Bmatrix} u \\ v \end{Bmatrix}, \quad s = 2, \\ \{\delta_k\}_m = \{u_k \quad v_k\}_m^T. \end{array}\right\} \qquad (1.17)$$

### 1.4.2. STRAINS

Once the displacement functions are known, it is possible to obtain the strains through appropriate differentiations with respect to the relevant coordinate variables $x$, $y$, or $z$, thus

$$\{\varepsilon\} = [B] \{\delta\}$$

$$= \sum_{m=1}^{r} [B]_m \{\delta\}_m$$

$$= \sum_{m=1}^{r} \sum_{k=1}^{s} [B_k]_m \{\delta_k\}_m . \qquad (1.18)$$

In this formulation, the so-called strains are really generalized strains and include normal and shear strain as well as bending and twisting curvatures. The $[B]$ matrix is referred to as the strain matrix.

For example, in the case of a bending strip, we have

$$\{\varepsilon\} = \begin{Bmatrix} \chi_x \\ \chi_y \\ 2\chi_{xy} \end{Bmatrix} = \begin{Bmatrix} -\dfrac{\partial^2 w}{\partial x^2} \\ -\dfrac{\partial^2 w}{\partial y^2} \\ 2\dfrac{\partial^2 w}{\partial x \, \partial y} \end{Bmatrix} . \qquad (1.19a)$$

Therefore

$$[B] = \begin{bmatrix} -\dfrac{\partial^2[N]}{\partial x^2} \\[2ex] -\dfrac{\partial^2[N]}{\partial y^2} \\[2ex] 2\dfrac{\partial^2[N]}{\partial x\,\partial y} \end{bmatrix} \tag{1.19b}$$

while for a two-dimensional plane strip,

$$\{\varepsilon\} = \begin{Bmatrix} \varepsilon_x \\ \varepsilon_y \\ \gamma_{xy} \end{Bmatrix} = \begin{Bmatrix} \dfrac{\partial u}{\partial x} \\[2ex] \dfrac{\partial v}{\partial y} \\[2ex] \dfrac{\partial u}{\partial y} + \dfrac{\partial v}{\partial x} \end{Bmatrix} \tag{1.20}$$

and the strain matrix $[B]$ can be written out in a similar manner.

### 1.4.3. STRESSES

The stresses are related to the strains by

$$\begin{aligned} \{\sigma\} &= [D]\{\varepsilon\} = [D][B]\{\delta\} \\ &= [D]\sum_{m=1}^{r}[B]_m\{\delta\}_m \\ &= [D]\sum_{m=1}^{r}\sum_{k=1}^{s}[B_k]_m\{\delta_k\}_m . \end{aligned} \tag{1.21}$$

The matrix $[D]$ is often referred to as the elasticity matrix or property matrix. For bending of an isotropic plate,

$$[D] = \frac{Et^3}{12(1-\nu^2)} \begin{bmatrix} 1 & \nu & 0 \\ \nu & 1 & 0 \\ 0 & 0 & \dfrac{1-\nu}{2} \end{bmatrix} \tag{1.22}$$

and for isotropic plane stress problems

$$[D] = \frac{E}{1-\nu^2} \begin{bmatrix} 1 & \nu & 0 \\ \nu & 1 & 0 \\ 0 & 0 & \dfrac{1-\nu}{2} \end{bmatrix} \quad (1.23)$$

Again, in the present context the term stresses represent generalized stresses including normal and shear stresses as well as bending and twisting moments and shearing forces.

### 1.4.4. MINIMIZATION OF TOTAL POTENTIAL ENERGY

#### (a) Strain energy

The strain energy of an elastic body is given by

$$U = \tfrac{1}{2} \int \{\varepsilon\}^T \{\sigma\} \, d(\text{vol.}). \quad (1.24a)$$

By virtue of (1.18) and (1.21), (1.24a) can be rewritten as

$$U = \tfrac{1}{2} \int \{\delta\}^T [B]^T [D] [B] \{\delta\} \, d(\text{vol.}). \quad (1.24b)$$

#### (b) Potential energy

The potential energy due to external surface loads $\{q\}$ can be written simply as

$$W = -\int \{f\}^T \{q\} \, d(\text{area}). \quad (1.25a)$$

Substituting (1.16) into (1.25a),

$$W = -\int \{\delta\}^T [N]^T \{q\} \, d(\text{area}). \quad (1.25b)$$

For a concentrated load, the above integral is reduced to the simple expression of load multiplied by corresponding displacement. For all other distributed loads the potential energy can be obtained through some simple integration process.

## (c) Total potential energy

The total potential energy, as stated previously, is the sum of the elastic strain energy stored in the body and the potential energy of the loads. Thus

$$\phi = U + W$$

$$= \frac{1}{2} \int \{\delta\}^T [B]^T [D] [B] \{\delta\} \, d(\text{vol.}) - \int \{\delta\}^T [N]^T \{q\} \, d(\text{area}). \quad (1.26)$$

## (d) Minimization procedure

The principle of minimum total potential energy requires that

$$\left\{ \frac{\partial \phi}{\partial \{\delta\}} \right\} = \{0\}. \quad (1.15)$$

Substituting (1.26) into (1.15) and performing the partial differentiation, we obtain

$$\left\{ \frac{\partial \phi}{\partial \{\delta\}} \right\} = \int [B]^T [D] [B] \{\delta\} \, d(\text{vol.}) - \int [N]^T \{q\} \, d(\text{area}) = \{0\} \quad (1.27)$$

or

$$[S] \{\delta\} - \{F\} = \{0\}, \quad (1.28)$$

in which

$$[S] = \int [B]^T [D] [B] \, d(\text{vol.})$$

$$= \int [[B]_1 [B]_2 \dots [B]_r]^T [D] [[B]_1 [B]_2 \dots [B]_r] \, d(\text{vol.})$$

$$= \int \begin{bmatrix} [B]_1^T [D] [B]_1 & [B]_1^T [D] [B]_2 & \dots & [B]_1^T [D] [B]_r \\ [B]_2^T [D] [B]_1 & [B]_2^T [D] [B]_2 & \dots & [B]_2^T [D] [B]_r \\ \dots & \dots & \dots & \dots \\ \dots & \dots & \dots & \dots \\ [B]_r^T [D] [B]_1 & [B]_r^T [D] [B]_2 & \dots & [B]_r^T [D] [B]_r \end{bmatrix} d(\text{vol.}) \quad (1.29a)$$

$$\text{or} \quad [S] = \begin{bmatrix} [S]_{11} & [S]_{12} & \cdots & [S]_{1r} \\ [S]_{21} & [S]_{22} & \cdots & [S]_{2r} \\ \cdots & \cdots & \cdots & \cdots \\ [S]_{r1} & [S]_{r2} & \cdots & [S]_{rr} \end{bmatrix} \quad (1.29b)$$

with
$$[S]_{mn} = \int [B]_m^T [D] [B]_n \, d(\text{vol.}) \quad (1.30)$$

Equation (1.30) can be further defined, in accordance with the total number of nodal lines $s$ within each strip, as

$$[S]_{mn} = \begin{bmatrix} [S_{11}] & [S_{12}] & \cdots & [S_{1s}] \\ [S_{21}] & [S_{22}] & \cdots & [S_{2s}] \\ \cdots & \cdots & \cdots & \cdots \\ [S_{s1}] & [S_{s2}] & \cdots & [S_{ss}] \end{bmatrix}_{mn} \quad (1.31)$$

in which the suffix $ij$ refers to nodal lines $i$ and $j$.

Finally, with the help of (1.18), we can write

$$[S_{ij}]_{mn} = \int [B_i]_m^T [D] [B_j]_n \, d(\text{vol.}). \quad (1.32)$$

It will be seen later that (1.32) yields the basic unit when the stiffness matrices of all the strips making up a structure are assembled to form an overall stiffness matrix.

The load matrix is interpreted from (1.27) and (1.28) to be

$$\{F\} = \int [N]^T \{q\} \, d(\text{area})$$

$$= \int \begin{bmatrix} [N]_1^T \\ [N]_2^T \\ \vdots \\ [N]_r^T \end{bmatrix} \{q\} \, d(\text{area}) \quad (1.33)$$

or for the $m$th term of the series only,

$$\{F\}_m = \int [N]_m^T \{q\} \, d(\text{area}). \quad (1.34)$$

Similar to what has been done in the development of the stiffness matrix, the load matrix can also be split up further by writing

$$\{F\}_m = \begin{Bmatrix} \{F_1\}_m \\ \{F_2\}_m \\ \vdots \\ \{F_s\}_m \end{Bmatrix} = \int \begin{Bmatrix} [N_1]_m^T \\ [N_2]_m^T \\ \vdots \\ \{N_s\}_m^T \end{Bmatrix} \{q\} \, d \,(\text{area}). \qquad (1.35)$$

Therefore    $\{F_i\}_m = \int [N_i]_m^T \{q\} \, d \,(\text{area}).$ \qquad (1.36)

A great deal of integration involving many different integrals are to be found in the expanded forms of (1.32) and (1.36). Although theoretically speaking all or nearly all of these integrals can be integrated analytically, the amount of manual work involved is very often of staggering proportions. The general rule is therefore that only those matrix coefficients using very simple integrals should be given in closed form solutions, while any integral involving some complexity should always be done numerically on the computer, using efficient integration rules such as the Gaussian quadrature formula.

## REFERENCES

1. O. C. ZIENKIEWICZ and Y. K. CHEUNG, *The Finite Element Method in Structural and Continuum Mechanics*, McGraw-Hill, 1967.
2. L. V. KANTOROVICH and V. I. KRYLOV, *Approximate Method of Higher Analysis*, Interscience Publishers, New York, 1958.
3. R. W. CLOUGH, The finite element method in structural mechanics, chapter 7 in *Stress Analysis* (ed. O. C. Zienkiewicz and G. S. Holister), John Wiley, 1965.
4. P. TONG and T. H. PIAN, The convergence of finite element method in solving linear elastic problems, *Int. J. Solids Struct.* **3**, 865–79 (1967).
5. V. VLAZOV, *General Theory of Shells and its Application in Engineering*, NASA TT F-99, April 1964.
6. O. C. ZIENKIEWICZ, B. M. IRONS, J. G. ERGATOUDIS, and F. C. SCOTT, Isoparametric and associated families for two and three dimensional analysis, Paper 13, *Finite Element Method in Stress Analysis* (ed. I. Holand and K. Bell), Techn. Univ. of Norway, Tapir Press, Norway, Trondheim, 1969.
7. R. L. TAYLOR, On completeness of shape functions for finite elements, *Int. J. Num. Meth. Eng.* **4**, 17–22 (1972).
8. W. D. MACDONALD and N. TRUFYN, *FORMAC Programmers Guide*, Institute of Computer Science, Univ. of Toronto, 1969.

# CHAPTER 2

# *Bending of plates and plate-beam systems with application to slab-beam bridges*

## 2.1. INTRODUCTION

The solutions presented herein are based on the classical plate theory assumptions in which certain approximations have been introduced so that a two-dimensional treatment becomes possible. A detailed account of the plate theory can be found in a number of tests by various authors.[1]

In bending analysis it is possible to formulate a strip in which each of the two opposite ends is either simply supported or clamped. Such a strip will then be used to analyse plates which are either simply supported, clamped free, or elastically supported by beams on the remaining two sides. The same computer program is therefore capable of analysing a wide range of plate problems.

The finite strip approach has considerable advantage over the conventional finite element method for the type of problems considered here in which the geometry is fairly simple and does not change in one direction, but, nevertheless, such structures are so frequently used in practice that a special and more economical treatment is warranted. The minimum number of DOF along a nodal line in this method is equal to twice (one deflection and one rotation) the number of terms used in the series, and this is usually considerably less than that for the finite element method, which requires a minimum of three times (one deflection and two rotations) the number of nodes along the same line. The size as well as the bandwidth of the matrix is greatly reduced, and consequently it can be handled by small computers or solved in a much shorter time. The economy is even more striking for the case of the simply supported strip (which is of great practical importance because of its use in the analysis of isotropic and orthotropic slab bridges), where a decoupling of the terms of the series occurs, and thus the size and bandwidth of the matrix is even more drastically reduced.

## 2.2. RECTANGULAR PLATE STRIP

The first paper on the finite strip method was presented by Cheung[2] on plate-bending problems using a simply supported rectangular strip. This was subsequently generalized by Cheung[3] to include other end conditions. A paper on the formulation of the simply supported rectangular strip was published independently at a later date by Powell and Ogden.[4]

In the above-mentioned publications, a lower order strip using a cubic polynomial function [shape function (b), (1.9b)] has been used throughout. Whilst the deflections obtained have been almost always very good, discontinuity of moments at strip boundaries and existence of some residual transverse moments at a free edge can be expected for coarse meshes. Of course, by increasing the number of strips used in the analysis such discontinuities and residual moments will diminish and even disappear altogether.

An alternative way to achieve higher accuracy without increasing the number of strips is to use higher order strips, and these can be formulated either by establishing higher order nodal line compatibilities [shape function (c), eqn. (1.9c)] or by introducing internal nodal lines [shape function (e), eqn. (1.9e)], in the same way as was done in finite element analysis.[5]

The use of shape function (c), which involves a quintic polynomial for a higher order bending strip, was originally suggested by Cheung,[6] although the actual formulation were carried out by Loo and Cusens,[7] who also worked out a second higher order strip using one additional internal nodal line.[8] The inclusion of end conditions, other than either of the two ends simply supported or clamped, was presented in a paper by Cheung and Cheung[9] for vibration analysis.

### 2.2.1. LOWER ORDER RECTANGULAR STRIP (LO2)

#### (a) Displacement function

Consider the strip shown in Fig. 2.1b. Each nodal line is free to move up and down in the $z$ direction and to rotate about the $y$ axis, with the result that there are two DOF per nodal line and a total of four DOF for the whole strip. A suitable displacement function can be written as

$$w = [N]\{\delta\} = \sum_{m=1}^{r} [N]_m \{\delta\}_m$$

$$= \sum_{m=1}^{r} Y_m [[C_1][C_2]] \{\delta\}_m \tag{2.1}$$

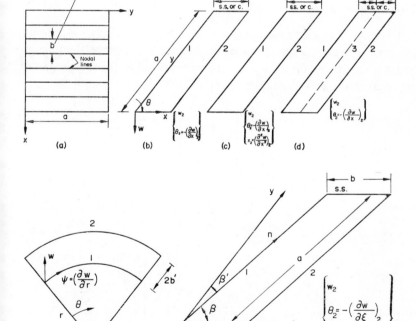

FIG. 2.1. (a) Top view of a structure divided into finite strips. (b) LO2 rectangular bending strip. (c) HO2 rectangular bending strip. (d) HO3 rectangular bending strip. (e) LO2 curved bending strip. (f) HO2 skew bending strip.

in which $[C_1]$, $[C_2]$ are given by (1.9b), and

$$\{\delta\}_m^T = \{w_{1m}\theta_{1m}w_{2m}\theta_{2m}\}^T$$

are the deflection and rotation parameters at the two longitudinal edges for the $m$th term of the series.

The above function assures compatibility of displacement values as well as its first partial derivatives at the interfaces of the strips, and therefore a convergent solution can be expected. For static analysis, (1.9a), (1.9b), and (1.9c) have been used as $Y_m$ (the series part of the displacement function) successfully, although for dynamic problems all six basic function series which appear in (1.9) have been included.

## (b) Strains (curvatures)

Once the displacement function is available, it is a relatively simple matter to obtain the curvatures by performing the appropriate differentiation. The strains for a plate in bending are given by the second partial derivatives of the displacement function, i.e.

$$\{\varepsilon\} = \begin{Bmatrix} -\chi_x \\ -\chi_y \\ 2\chi_{xy} \end{Bmatrix} = \begin{Bmatrix} -\dfrac{\partial^2 w}{\partial x^2} \\[2mm] -\dfrac{\partial^2 w}{\partial y^2} \\[2mm] 2\dfrac{\partial^2 w}{\partial x \partial y} \end{Bmatrix}$$

$$= \sum_{m=1}^{r} \begin{Bmatrix} -\dfrac{\partial^2 [N]_m}{\partial x^2} \\[2mm] -\dfrac{\partial^2 [N]_m}{\partial y^2} \\[2mm] 2\dfrac{\partial^2 [N]_m}{\partial x \partial y} \end{Bmatrix} \{\delta\}_m$$

$$= \sum_{m=1}^{r} [B]_m \{\delta\}_m . \tag{2.2}$$

The matrix $[B]_m$ is obtained through the appropriate differentiations.

$$
= \begin{bmatrix}
\dfrac{6}{b_2}(1-2\bar{x})Y_m & \dfrac{2}{b}(2-3\bar{x})Y_m & \dfrac{6}{b^2}(-1+2\bar{x})Y_m & \dfrac{2}{b}(-3\bar{x}+1)Y_m \\[2mm]
-(1-3\bar{x}^2+2\bar{x}^3)Y_m'' & -x(1-2\bar{x}+\bar{x}^2)Y_m'' & -(3\bar{x}^2-2\bar{x}^3)Y_m'' & -x(\bar{x}^2-\bar{x})Y_m'' \\[2mm]
\dfrac{2}{b}(-6\bar{x}+6\bar{x}^2)Y_m' & 2(1-4\bar{x}+3\bar{x}^2)Y_m' & \dfrac{2}{b}(6\bar{x}-6\bar{x}^2)Y_m' & 2(3\bar{x}^2-2\bar{x})Y_m'
\end{bmatrix}
$$

$$(2.3)$$

### (c) Stresses (moments)

The moments are related to the strains through the material properties of the strip. In the present formulation the more general case of ortho-tropic properties will be assumed. Therefore

$$
\{\sigma\} = \begin{Bmatrix} M_x \\ M_y \\ M_{xy} \end{Bmatrix} = [D]\{\varepsilon\}
$$

$$
= [D] \sum_{m=1}^{r} [B]_m \{\delta\}_m \tag{2.4}
$$

in which the property matrix is

$$
[D] = \begin{bmatrix} D_x & D_1 & 0 \\ D_1 & D_y & 0 \\ 0 & 0 & D_{xy} \end{bmatrix} \tag{2.5}
$$

and $D_x$, $D_y$, $D_1$ and $D_{xy}$ are the orthotropic plate constants[1] given by

$$
D_x = \frac{E_x t^3}{12(1-\nu_x\nu_y)},
$$

$$
D_y = \frac{E_y t^3}{12(1-\nu_x\nu_y)},
$$

$$
D_1 = \frac{\nu_x E_y t^3}{12(1-\nu_x\nu_y)} = \frac{\nu_y E_x t^3}{12(1-\nu_x\nu_y)},
$$

$$
D_{xy} = \frac{G t^3}{12}.
$$

In the above expressions, $E_x$, $E_y$, $v_x$, $v_y$, and $G$ are the elastic constants while $t$ is the thickness of the plate strip. For an isotropic plate, it is only necessary to set $E_x = E_y = E$, $v_x = v_y$, and $G = \dfrac{E}{2(1+v)}$.

*(d) Stiffness matrix*

The stiffness matrix can now be derived according to (1.29) and (1.30). For plate-bending problems, in which the integration in the thickness direction has already been carried out, (1.30) can be modified as

$$[S]_{mn} = \int [B]_m^T [D] [B]_n \, d\,(\text{area}). \tag{2.6}$$

An explicit form of the stiffness matrix $[S]_{mn}$ is listed in Table 2.1.

For the special case of a simply supported strip, we have, from (1.6a),

$$Y_m = \sin \frac{m\pi y}{a} = \sin k_m y.$$

Therefore

$$Y_m' = k_m \cos k_m y \quad \text{and} \quad Y_m'' = -k_m^2 \sin k_m y.$$

Consequently all the integrals $I_1$ to $I_5$ and also $[S]_{mn}$ of Table 2.1 are equal to zero for $m \neq n$. This can be proved with the help of (1.7) as follows:

$$I_1 = \int_0^a Y_m Y_n \, dy = 0,$$

$$I_4 = \int_0^a Y_m'' Y_n'' \, dy = 0,$$

$$I_2 = \int_0^a Y_m'' Y_n \, dy = k_m^2 \int_0^a \sin k_m y \, \sin k_n y \, dy = 0,$$

$$I_3 = \int_0^a Y_m Y_n'' \, dy = -k_n^2 \int_0^a \sin k_m y \, \sin k_n y \, dy = 0,$$

$$I_5 = \int_0^a Y_m' Y_n' \, dy = k_m k_n \int_0^a \cos k_m y \, \cos k_n y \, dy = 0,$$

where $k_m = m\pi/a$ and $k_n = n\pi/a$ respectively.

Because of this orthogonal property, (1.29b) now takes up the form

$$[S] = \begin{bmatrix} [S]_{11} & & & 0 \\ & [S]_{22} & & \\ & & \ddots & \\ 0 & & & [S]_{rr} \end{bmatrix} \tag{2.7}$$

TABLE 2.1. BENDING STIFFNESS MATRIX FOR A RECTANGULAR STRIP WITH ANY END CONDITIONS

$$[S]_{mn} = \frac{1}{420b^3}$$

| | | | |
|---|---|---|---|
| 5040 $D_x I_1$<br>−504 $b^2 D_1 I_2$<br>−504 $b^2 D_1 I_3$<br>156 $b^4 D_y I_4$<br>2016 $b^2 D_{xy} I_5$ | 2520 $b\ D_x I_1$<br>−462 $b^3 D_1 I_2$<br>−42 $b^3 D_1 I_3$<br>22 $b^5 D_y I_4$<br>168 $b^3 D_{xy} I_5$ | −5040 $D_x I_1$<br>504 $b^2 D_1 I_2$<br>504 $b^2 D_1 I_3$<br>54 $b^4 D_y I_4$<br>−2016 $b^2 D_{xy} I_5$ | 2520 $b\ D_x I_1$<br>−42 $b^3 D_1 I_2$<br>−42 $b^3 D_1 I_3$<br>−13 $b^5 D_y I_4$<br>168 $b^3 D_{xy} I_5$ |
| 2520 $b\ D_x I_1$<br>−462 $b^3 D_1 I_2$<br>−42 $b^3 D_1 I_3$<br>22 $b^5 D_y I_4$<br>168 $b^3 D_{xy} I_5$ | 1680 $b^2 D_x I_1$<br>−56 $b^4 D_1 I_2$<br>−56 $b^4 D_1 I_3$<br>4 $b^6 D_y I_4$<br>224 $b^4 D_{xy} I_5$ | −2520 $b\ D_x I_1$<br>42 $b^3 D_1 I_2$<br>42 $b^3 D_1 I_3$<br>13 $b^5 D_y I_4$<br>−168 $b^3 D_{xy} I_5$ | 840 $b^2 D_x I_1$<br>14 $b^4 D_1 I_2$<br>14 $b^4 D_1 I_3$<br>−3 $b^6 D_y I_4$<br>−56 $b^4 D_{xy} I_5$ |
| −5040 $D_x I_1$<br>504 $b^2 D_1 I_3$<br>504 $b^2 D_1 I_2$<br>54 $b^4 D_y I_4$<br>−2016 $b^2 D_{xy} I_5$ | −2520 $b\ D_x I_1$<br>42 $b^3 D_1 I_3$<br>42 $b^3 D_1 I_2$<br>13 $b^5 D_y I_4$<br>−168 $b^3 D_{xy} I_5$ | 5040 $D_x I_1$<br>−504 $b^2 D_1 I_2$<br>−504 $b^2 D_1 I_3$<br>156 $b^4 D_y I_4$<br>2016 $b^2 D_{xy} I_5$ | −2520 $b\ D_x I_1$<br>462 $b^3 D_1 I_2$<br>42 $b^3 D_1 I_3$<br>−22 $b^5 D_y I_4$<br>−168 $b^3 D_{xy} I_5$ |
| 2520 $b\ D_x I_1$<br>−42 $b^3 D_1 I_3$<br>−42 $b^3 D_1 I_2$<br>−13 $b^5 D_y I_4$<br>168 $b^3 D_{xy} I_5$ | 840 $b^2 D_x I_1$<br>14 $b^4 D_1 I_3$<br>14 $b^4 D_1 I_2$<br>−3 $b^6 D_y I_4$<br>−56 $b^4 D_{xy} I_5$ | −2520 $b\ D_x I_1$<br>462 $b^3 D_1 I_2$<br>42 $b^3 D_1 I_3$<br>−22 $b^5 D_y I_4$<br>−168 $b^3 D_{xy} I_5$ | 1680 $b^2 D_x I_1$<br>−56 $b^4 D_1 I_2$<br>−56 $b^4 D_1 I_3$<br>4 $b^6 D_y I_4$<br>224 $b^4 D_{xy} I_5$ |

$I_1 = \int_0^a Y_m Y_n\,dy$; $I_2 = \int_0^a Y_m'' Y_n\,dy$; $I_3 = \int_0^a Y_m Y_n''\,dy$; $I_4 = \int_0^a Y_m'' Y_n''\,dy$; $I_5 = \int_0^a Y_m' Y_n'\,dy$;

(for $m \neq n$, $I_1 = I_4 = 0$).

and all the terms of the series are thus uncoupled. The final form of the stiffness matrix $[S]_{mn}$ for a simply supported strip is given in Table 2.2.

A closer examination of Table 2.2 will reveal a property peculiar of the finite strip method. In standard finite element formulation, one way of checking the correctness of the element stiffness matrix is to sum all the stiffness coefficients relevant to each nodal force component (e.g. nodal shear forces) that appears in the same column of the matrix and to make sure that such sums do vanish. Because the element is entirely unsupported, the fact that such sums are equal to zero would represent equilibrium states for the various force components.

However, such a method of checking is no longer correct when applied to the stiffness matrix of a strip. This is due to the fact that support reactions exist in most cases, and although they are not included in the stiffness matrix they nevertheless contribute towards the equilibrium of the strip.

## (e) Consistent load matrix

The load matrix is given by (1.35) as

$$\{F\}_m = \int_0^a \int_0^b \begin{Bmatrix} [N_1]_m^T \\ [N_2]_m^T \end{Bmatrix} \{q\}\, dx\, dy. \tag{2.8a}$$

For uniformly distributed load

$$\{F\}_m = q \int_0^b \begin{Bmatrix} [C_1]^T \\ [C_2]^T \end{Bmatrix} dx \int_0^a Y_m\, dy$$

$$= q \begin{Bmatrix} \dfrac{b}{2} \\[2mm] \dfrac{b^2}{12} \\[2mm] \dfrac{b}{2} \\[2mm] \dfrac{-b^2}{12} \end{Bmatrix} \int_0^a Y_m\, dy \tag{2.8b}$$

in which $[C_1]$ and $[C_2]$ are, as before, given by (1.9b).

For a concentrated load $P_c$ at a point $x = x_c$, $y = y_c$, it is only necessary to substitute the coordinates of the load into the first part of (2.8b) and

TABLE 2.2. STIFFNESS MATRIX OF SIMPLY SUPPORTED RECTANGULAR STRIP

$$[S]_{mn} =
\begin{bmatrix}
\dfrac{13ab}{70}k_m^4 D_y + \dfrac{12a}{5b}k_m^2 D_{xy} + \dfrac{6a}{5b}k_m^2 D_1 + \dfrac{6a}{b^3}D_x & & & \\[2ex]
\dfrac{3a}{5}k_m^2 D_1 + \dfrac{a}{5}k_m^2 D_{xy} + \dfrac{3a}{b^2}D_x + \dfrac{11ab^2}{420}k_m^4 D_y & \dfrac{ab^3}{210}k_m^4 D_y + \dfrac{4ab}{15}k_m^2 D_{xy} + \dfrac{2ab}{15}k_m^2 D_1 + \dfrac{2a}{b}D_x & & \text{Symmetrical} \\[2ex]
\dfrac{9ab}{140}k_m^4 D_y - \dfrac{12a}{5b}k_m^2 D_{xy} - \dfrac{6a}{5b}k_m^2 D_1 - \dfrac{6a}{b^3}D_x & \dfrac{13ab^2}{840}k_m^4 D_y - \dfrac{a}{5}k_m^2 D_{xy} - \dfrac{a}{10}k_m^2 D_1 - \dfrac{3a}{b^2}D_x & \dfrac{13ab}{70}k_m^4 D_y + \dfrac{12a}{5b}k_m^2 D_{xy} + \dfrac{6a}{5b}k_m^2 D_1 + \dfrac{6a}{b^3}D_x & \\[2ex]
-\dfrac{13ab^2}{840}k_m^4 D_y + \dfrac{a}{5}k_m^2 D_{xy} + \dfrac{a}{10}k_m^2 D_1 + \dfrac{3a}{b^2}D_x & -\dfrac{3ab^3}{840}k_m^4 D_y - \dfrac{ab}{15}k_m^2 D_{xy} - \dfrac{ab}{30}k_m^2 D_1 + \dfrac{a}{b}D_x & -\dfrac{11ab^2}{420}k_m^4 D_y - \dfrac{a}{5}k_m^2 D_{xy} - \dfrac{3a}{5}k_m^2 D_1 - \dfrac{3a}{b^2}D_x & \dfrac{ab^3}{210}k_m^4 D_y + \dfrac{4ab}{15}k_m^2 D_{xy} + \dfrac{2ab}{15}k_m^2 D_1 + \dfrac{2a}{b}D_x
\end{bmatrix}$$

$$k_m = \frac{m\pi}{a}$$

also to dispense with the integration. Letting $x_c/b = \bar{x}_c$, we have

$$\{F\}_m = P_c Y_m(y_c) \begin{Bmatrix} (1-3\bar{x}_c^2+2\bar{x}_c^3) \\ x_c(1-2\bar{x}_c+\bar{x}_c^2) \\ (3\bar{x}_c^2-2\bar{x}_c^3) \\ x_c(\bar{x}_c^2-\bar{x}_c) \end{Bmatrix}. \tag{2.8c}$$

Since the distribution of the transverse moment is linear from one edge to the other of the strip, it would not be possible to obtain the maximum moment under the point load. Therefore, if circumstances permit, it is more desirable to actually insert a nodal line under the point load.

### 2.2.2. HIGHER ORDER RECTANGULAR STRIP WITH TWO NODAL LINES (HO2)

In this strip, the transverse curvature amplitude is also used as a nodal displacement parameter in addition to the standard deflection and rotation variables. As a result, continuous curvatures and consequently moments will now exist at the interface between neighbouring strips, and therefore more accurate results can be expected when compared with a lower order strip analysis for the same number of strips. Two points must be borne in mind, however. The first point is that due to the increase in nodal displacement parameters the half bandwidth of the final stiffness matrix is increased by 50%. This is not really serious, since it has been observed from experience that the increase in accuracy more than compensates the increase in computation effort. The second point is that such a strip cannot be used, at least not in a straightforward manner, for plates with abrupt changes in thickness or material properties in the transverse direction because at the line where such abrupt change occurs, the curvatures should always be discontinuous. The nature of this disadvantage is more serious since the scope of application of such higher order strips will be more or less limited to constant thickness plates.

The nodal displacement parameters for the strip shown in Fig. 2.1c include the displacement and its first and second partial derivatives with respect to $x$. A suitable displacement function can be obtained as a product of the shape function (c) given by (1.9c) and the basic function series as follows:

$$w = \sum_{m=1}^{r} [N]_m \{\delta\}_m = \sum_{m=1}^{r} Y_m [[C_1] [C_2]]_m \{\delta\}_m, \qquad (2.9a)$$

where $[C_1]$ and $[C_2]$ are now given by (1.9c)

and

$$\{\delta\}_m = [w_{1m}\theta_{1m}\chi_{1m}w_{2m}\theta_{2m}\chi_{2m}]^T \qquad (2.9b)$$

in which $\chi_{im} = \left(\dfrac{\partial^2 w}{\partial x^2}\right)_{im}$ is the curvature parameter at nodal line $i$ for the $m$th term of the series.

With the displacement function established, the procedure outlined in Section 1.3 can be repeated in the same way as for the lower order strip, and the stiffness and load matrices obtained. The general form of $[S]_{mn}$ and $\{F\}_m$ has been worked out by Cheung[10] but will not be given here because of its complexity. However, for a simply supported strip, $[S]_{mm}$ and $\{F\}_m$ are quite simple and are listed in Tables 2.3 and 2.4.

### 2.2.3. HIGHER ORDER STRIP (HO3) WITH ONE INTERNAL NODAL LINE (FIG. 2.1d)

This strip uses an additional internal nodal line which, for the sake of convenience, is usually placed midway between the two longitudinal edges. Since each nodal line includes two displacement parameters, a deflection and a rotation, it is apparent that the half bandwidth of the final stiffness matrix is also 50% higher than that for a lower order strip. However, due to the fact that the internal nodal displacement parameters are not connected with anything else apart from those of the same strip, it is possible to eliminate these two variables before assembly through the process of static condensation, which will be described in the following paragraphs.

Let the stiffness equations of a strip be partitioned and the subscripts $i$ and $o$ be used to represent the inner and outer nodal lines respectively. Then

$$\begin{bmatrix} [S_{oo}] & [S_{oi}] \\ [S_{oi}]^T & [S_{ii}] \end{bmatrix} \begin{Bmatrix} \{\delta_o\} \\ \{\delta_i\} \end{Bmatrix} = \begin{Bmatrix} \{F_o\} \\ \{F_i\} \end{Bmatrix}. \qquad (2.10)$$

TABLE 2.3. STIFFNESS MATRIX $[S]_{mm}$ FOR A BENDING SIMPLY SUPPORTED STRIP WITH CURVATURE COMPATIBILITY (LOO[11])

$$
\begin{bmatrix}
S1 & & & & & \\
S4 & S2 & & \text{Symmetrical} & & \\
S5 & S6 & S3 & & & \\
S10 & S7 & S8 & S1 & & \\
-S7 & S11 & S9 & -S4 & S2 & \\
S8 & -S9 & S12 & S5 & -S6 & S3
\end{bmatrix}
$$

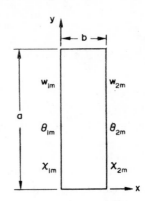

where,

$$S1 = \frac{120B}{7b^3} + \frac{20C}{7b} + \frac{181bD}{462} + \frac{10E}{7b}$$

$$S2 = \frac{192B}{35b} + \frac{16bC}{35} + \frac{52b^3D}{3465} + \frac{8bE}{35}$$

$$S3 = \frac{3bB}{35} + \frac{b^3C}{315} + \frac{b^5D}{9240} + \frac{b^3E}{630}$$

$$S4 = \frac{60B}{7b^2} + \frac{10C}{7} + \frac{311b^2D}{4620} + \frac{3E}{14}$$

$$S5 = \frac{3B}{7b} + \frac{bC}{42} + \frac{281b^3D}{55\,440} + \frac{bE}{84}$$

$$S6 = \frac{11B}{35} + \frac{b^2C}{30} + \frac{23b^4D}{18\,480} + \frac{b^2E}{60}$$

$$S7 = \frac{-60B}{7b^2} - \frac{3C}{7} + \frac{151b^2D}{4620} - \frac{3E}{14}$$

$$S8 = \frac{-3B}{7b} - \frac{bC}{42} + \frac{181b^3D}{55\,440} - \frac{bE}{84}$$

$$S9 = \frac{4B}{35} - \frac{b^2C}{105} - \frac{13b^4D}{13\,860} - \frac{b^2E}{210}$$

$$S10 = \frac{-120B}{7b^3} - \frac{20C}{7b} + \frac{25bD}{231} - \frac{10E}{7b}$$

$$S11 = \frac{108B}{35b} - \frac{bC}{35} - \frac{19b^3D}{1980} - \frac{bE}{70}$$

$$S12 = \frac{bB}{70} + \frac{b^3C}{630} + \frac{b^5D}{11\,088} + \frac{b^3E}{1260}$$

in which,

$$B = \frac{aD_x}{2} \qquad D = \frac{ak_m^4 D_y}{2}$$

$$C = \frac{ak_m^2 D_1}{2} \qquad E = 2ak_m^2 D_{xy}$$

$$k_m = \frac{m\pi}{a}$$

TABLE 2.4. LOAD MATRICES $\{F\}_m$ DUE TO APPLIED LOADS FOR A BENDING
SIMPLY SUPPORTED STRIP WITH CURVATURE COMPATIBILITY (LOO[11])

(a) Concentrated load

$$\begin{bmatrix} 1 - \dfrac{10x_e^3}{b^3} + \dfrac{15x_e^4}{b^4} - \dfrac{6x_e^5}{b^5} \\[2ex] x_e - \dfrac{6x_e^3}{b^2} + \dfrac{8x_e^4}{b^3} - \dfrac{3x_e^5}{b^4} \\[2ex] \dfrac{x_e^2}{2} - \dfrac{1.5x_e^3}{b} + \dfrac{1.5x_e^4}{b^2} - \dfrac{x_e^6}{2b^2} \\[2ex] \dfrac{10x_e^3}{b^3} - \dfrac{15x_e^4}{b^4} + \dfrac{6x_e^5}{b^5} \\[2ex] -\dfrac{4x_e^3}{b^2} + \dfrac{7x_e^4}{b^3} - \dfrac{3x_e^5}{b^4} \\[2ex] \dfrac{x_e^3}{2b} - \dfrac{x_e^4}{b^2} + \dfrac{x_e^5}{2b^3} \end{bmatrix} P_e \sin k_m y_e$$

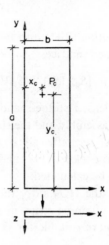

(b) Patch load

$$\begin{bmatrix} x_0 - \dfrac{2.5x_0^4}{b^3} + \dfrac{3x_0^5}{b^4} - \dfrac{x_0^6}{b^5} \\[2ex] \dfrac{x_0^2}{2} - \dfrac{1.5x_0^4}{b^2} + \dfrac{1.6x_0^5}{b^3} - \dfrac{x_0^6}{2b^4} \\[2ex] \dfrac{x_0^3}{6} - \dfrac{3x_0^4}{8b} + \dfrac{0.3x_0^5}{b^2} - \dfrac{x_0^6}{12b^3} \\[2ex] \dfrac{2.5x_0^4}{b^3} - \dfrac{3x_0^5}{b^4} + \dfrac{x_0^6}{b^5} \\[2ex] -\dfrac{x_0^4}{b^2} + \dfrac{1.4x_0^5}{b^3} - \dfrac{x_0^6}{2b^4} \\[2ex] \dfrac{x_0^4}{8b} - \dfrac{x_0^5}{5b^2} + \dfrac{x_0^6}{12b^3} \end{bmatrix} Q_0 C_m$$

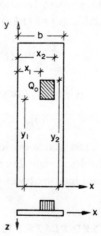

where $x_0^n = x_2^n - x_1^n$ and $C_m = \dfrac{1}{k_m} (\cos k_m y_1 - \cos k_m y_2)$

The lower half of (2.10) can be rearranged as

$$[S_{ii}]\{\delta_i\} = \{F_i\} - [S_{oi}]^T\{\delta_o\}$$

or
$$\{\delta_i\} = [S_{ii}]^{-1}(\{F_i\} - [S_{oi}]^T\{\delta_o\}). \tag{2.11}$$

Substituting (2.11) into the upper half of (2.10), we finally arrived at the modified matrices

$$([S_{oo}] - [S_{oi}][S_{ii}]^{-1}[S_{oi}]^T\{\delta_o\} = (\{F_o\} - [S_{oi}][S_{ii}]^{-1}\{F_i\}). \tag{2.12}$$

The condensed stiffness and load matrices are now used as the basic units for assembly and the size is now identical to that of a lower order strip. Of course, after all the $\{\delta_o\}$ have been computed, it would be necessary to go back to (2.11) and compute for $\{\delta_i\}$ before any stresses of the strip can be calculated.

Unlike the HO$_2$ strip, there is no curvature continuity at the interfaces of adjoining strips in the present formulation, and therefore this strip is more suitable for general plate problems with abrupt property change.

The displacement function used in this case is simply the product of shape function (e) of (1.9e) and the basic function series

$$w = \sum_{m=1}^{r}[N]_m\{\delta\}_m = \sum_{m=1}^{r}Y_m[[C_1][C_2][C_3]]\{\delta\}_m, \tag{2.13a}$$

where $[C_1]$, $[C_2]$, $[C_3]$ are given by (1.9e)

and
$$\{\delta\}_m = [w_{1m}\theta_{1m}w_{2m}\theta_{2m}w_{3m}\theta_{3m}]^T. \tag{2.13b}$$

The stiffness and some load matrices for a simply supported strip are given in Tables 2.5 and 2.6. Some further details concerning the condensed stiffness and load matrices can be found elsewhere[11].

## 2.3. CURVED PLATE STRIP[12]

In modern construction, many structures are curved in plan because of aesthetical or functional reasons, and sector plates bounded by two radial lines and one or two (concentric) circular arcs are especially popular. These curved structures are usually either plain curved slabs or are made up of a system of slabs and multiple stiffening girders. These

TABLE 2.5. STIFFNESS MATRIX FOR A HO3 BENDING STRIP (LOO[11])

$$\begin{bmatrix} S1 \\ S2 & S7 & & \text{Symmetrical} \\ S3 & S8 & S11 \\ S4 & S9 & 0 & S12 \\ S5 & -S6 & S3 & -S4 & S1 \\ S6 & S10 & -S8 & S9 & -S2 & S7 \end{bmatrix}$$

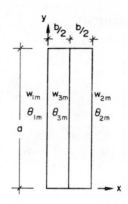

where,

$$S1 = \frac{5092B}{35b^3} + \frac{278C}{105b} + \frac{523bD}{3465} + \frac{278E}{105b}$$

$$S2 = \frac{1138B}{35b^2} + \frac{59C}{105} - \frac{19b^2D}{2310} + \frac{13E}{210}$$

$$S3 = \frac{-512B}{5b^3} - \frac{256C}{105b} + \frac{4bD}{63} - \frac{256E}{105b}$$

$$S4 = \frac{384B}{7b^2} + \frac{8C}{21} - \frac{8b^2D}{693} + \frac{8E}{21}$$

$$S5 = \frac{-1508B}{35b^3} - \frac{22C}{105b} + \frac{131bD}{6930} - \frac{22E}{105b}$$

$$S6 = \frac{242B}{35b^2} - \frac{C}{70} - \frac{29b^2D}{13\,860} - \frac{E}{70}$$

$$S7 = \frac{332B}{35b} + \frac{2bC}{45} + \frac{2b^3D}{3465} + \frac{2bE}{45}$$

$$S8 = \frac{-128B}{5b^2} - \frac{8C}{105} + \frac{2b^2D}{315} - \frac{8E}{105}$$

$$S9 = \frac{64B}{7b} - \frac{4bC}{315} - \frac{b^3D}{1155} - \frac{4bE}{315}$$

$$S10 = \frac{38B}{35b} - \frac{bC}{126} - \frac{b^3D}{4620} - \frac{bE}{126}$$

$$S11 = \frac{1024B}{5b^3} + \frac{512C}{105b} + \frac{128bD}{315} + \frac{512E}{10\,5b}$$

$$S12 = \frac{256B}{7b} + \frac{128bC}{315} + \frac{32b^3D}{3465} + \frac{128bE}{315}$$

in which,

$$B = \frac{aD_x}{2} \qquad\qquad D = \frac{ak_m^4 D_y}{2}$$

$$C = ak_m^2 D_1 \qquad\qquad E = 2ak_m^2 D_{xy}$$

$$k_m = \frac{m\pi}{a}$$

TABLE 2.6. LOAD MATRICES DUE TO APPLIED LOADS FOR A HO3 STRIP (Loo[11])

(a) Concentrated load

$$\{F\}_m = \begin{bmatrix} 1 - \dfrac{23x_c^2}{b^2} + \dfrac{66x_c^3}{b^3} - \dfrac{68x_c^4}{b^4} - \dfrac{24x_c^5}{b^5} \\[2ex] x_c - \dfrac{6x_c^2}{b} + \dfrac{13x_c^3}{b^2} - \dfrac{12x_c^4}{b^3} + \dfrac{4x_c^5}{b^4} \\[2ex] \dfrac{16x_c^2}{b^2} - \dfrac{32x_c^3}{b^3} + \dfrac{16x_c^4}{b^4} \\[2ex] -\dfrac{8x_c^2}{b} + \dfrac{32x_c^3}{b^2} - \dfrac{40x_c^4}{b^3} + \dfrac{16x_c^5}{b^4} \\[2ex] \dfrac{7x_c^2}{b^2} - \dfrac{34x_c^3}{b^3} + \dfrac{52x_c^4}{b^4} - \dfrac{24x_c^5}{b^5} \\[2ex] -\dfrac{x_c^2}{b} + \dfrac{5x_c^3}{b^2} - \dfrac{8x_c^4}{b^3} + \dfrac{4x_c^5}{b^4} \end{bmatrix} P_c \sin k_m y_c$$

(b) Patch load

$$\{F\}_m = \begin{bmatrix} x_0 - \dfrac{23x_0^3}{3b^2} + \dfrac{33x_0^4}{2b^3} + \dfrac{68x_0^5}{5b^4} + \dfrac{4x_0^6}{b^5} \\[2ex] \dfrac{x_0^2}{2} - \dfrac{2x_0^3}{b} + \dfrac{13x_0^4}{4b^2} - \dfrac{12x_0^5}{5b^3} + \dfrac{2x_0^6}{3b^4} \\[2ex] \dfrac{16x_0^3}{3b^2} - \dfrac{8x_0^4}{b^3} + \dfrac{16x_0^5}{5b^4} \\[2ex] -\dfrac{8x_0^3}{3b} + \dfrac{8x_0^4}{b^2} - \dfrac{8x_0^5}{b^3} + \dfrac{8x_0^6}{3b^4} \\[2ex] \dfrac{7x_0^3}{3b^2} - \dfrac{17x_0^4}{2b^3} + \dfrac{52x_0^5}{5b^4} - \dfrac{4x_0^6}{b^5} \\[2ex] -\dfrac{x_0^3}{3b} + \dfrac{5x_0^4}{4b^2} - \dfrac{8x_0^5}{5b^3} + \dfrac{2x_0^6}{3b^4} \end{bmatrix} Q_0 C_m$$

in which,

$$C_m = (\cos k_m y_1 - \cos k_m y_2)/k_m \quad \text{and} \quad x_0^n = x_2^n - x_1^n$$

multiple girders are usually arranged orthogonally along the radii and arcs of concentric circles, and the entire system can be treated as an equivalent cylindrical orthotropic plate. The research conducted so far is mostly on curved beam assemblages or on constant thickness isotropic plates.[13, 14]

In the previous discussions, rectangular strips using a Cartesian coordinate system have been developed and applied successfully to the analysis of rectangular orthotropic plates. It is immediately apparent that the same approach can be applied to the analysis of cylindrical orthotropic curved plates by using curved strips and a polar coordinate system. In fact it will be demonstrated that any rectangular plate can be analysed by the curved strip program by simply adopting a very small subtended angle together with a large radius.

The curved strip which will be presented here is a lower order strip. Higher order strips can also be formulated without any difficulty if desired.

### (a) Displacement function

A comparison of the curved strip shown in Fig. 2.1e and its counterpart in a Cartesian coordinate system (Fig. 2.1b) will show clearly the various terms which correspond to one another in the two coordinate systems. Such a comparison is listed in Table 2.7, and from these data the displacement function for the curved strip is readily deduced.

The displacement function can be further simplified by introducing the dimensionless variables

$$b' = \frac{r_2 - r_1}{2} \quad \text{and} \quad R = \frac{r - r_1}{b'}, \quad \text{so that at} \quad r = r_1, \quad R = 0,$$

$$\text{and at} \quad r = r_2, \quad R = 2.$$

Thus, finally,

$$w = \sum_{m=1}^{r} \Theta_m \left[ \left( 1 - \frac{3}{4} R^2 + \frac{1}{4} R^3 \right), \quad b' \left( R - R^2 + \frac{R^3}{4} \right), \right.$$

$$\left. \left( \frac{3}{4} R^2 - \frac{1}{4} R^3 \right), \quad b' \left( \frac{R^3}{4} - \frac{R^2}{2} \right) \right] \begin{Bmatrix} w_{1m} \\ \psi_{1m} \\ w_{2m} \\ \psi_{2m} \end{Bmatrix}$$

$$= \sum_{m=1}^{r} [N]_m \{\delta\}_m. \tag{2.14}$$

TABLE 2.7. CONVERSION OF DISPLACEMENT FUNCTION FOR BENDING STRIP

**Rectangular strip**

$$w = \sum_{m=1}^{r} \left[ \left( \left(1 - \frac{3x^2}{b^2} + \frac{2x^3}{b^3}\right),\ \left(x - \frac{2x^2}{b} + \frac{x^3}{b^2}\right),\ \left(\frac{3x^2}{b^2} - \frac{2x^3}{b^3}\right),\ \left(\frac{x^3}{b^2} - \frac{x^2}{b}\right) \right) Y_m [w_{1m}\,\theta_{1m}\,w_{2m}\,\theta_{2m}]^T \right]$$

**Curved strip**

$$w = \sum_{m=1}^{r} \left[ \left( \left(1 - 3\left(\frac{r-r_1}{r_2-r_1}\right)^2 + 2\left(\frac{r-r_1}{r_2-r_1}\right)^3\right),\ \left((r-r_1) - \frac{2(r-r_1)^2}{r_2-r_1} + \frac{(r-r_1)^3}{(r_2-r_1)^2}\right),\ \left(3\left(\frac{r-r_1}{r_2-r_1}\right)^2 - 2\left(\frac{r-r_1}{r_2-r_1}\right)^3\right),\ \left(\frac{(r-r_1)^3}{(r_2-r_1)^2} - \frac{(r-r_1)^2}{(r_2-r_1)}\right) \right) \Theta_m [w_{1m}\,\psi_{1m}\,w_{2m}\,\psi_{2m}]^T \right]$$

**Conversion**

| $x$ | $y$ | $a$ | $b$ | $Y_m$ | $\theta = \dfrac{\partial w}{\partial x}$ |
|-----|-----|-----|-----|-------|-------------------------------------------|
| $r - r_1$ | $\theta$ | $\alpha$ | $r_2 - r_1$ | $\Theta_m$ | $\psi = \dfrac{\partial w}{\partial r}$ |

## (b) Stiffness and load matrices

The curvatures of a plate in polar coordinates are given by[1]

$$\{\varepsilon\} = \begin{Bmatrix} -\chi_r \\ -\chi_\theta \\ 2\chi_{r\theta} \end{Bmatrix} = \begin{bmatrix} -\dfrac{\partial^2 w}{\partial r^2} \\ -\dfrac{1}{r}\left(\dfrac{1}{r}\dfrac{\partial^2 w}{\partial \theta^2} + \dfrac{\partial w}{\partial r}\right) \\ -\dfrac{2}{r}\left(\dfrac{\partial^2 w}{\partial r\,\partial \theta} - \dfrac{1}{r}\dfrac{\partial w}{\partial \theta}\right) \end{bmatrix}$$

$$= \sum_{m=1}^{r} \begin{bmatrix} -\dfrac{\partial^2 [N]_m}{\partial r^2} \\ -\dfrac{1}{r}\left(\dfrac{1}{r}\dfrac{\partial^2 [N]_m}{\partial \theta^2} + \dfrac{\partial [N]_m}{\partial r}\right) \\ -\dfrac{2}{r}\left(\dfrac{\partial^2 [N]_m}{\partial r\,\partial \theta} - \dfrac{1}{r}\dfrac{\partial [N]_m}{\partial \theta}\right) \end{bmatrix} \{\delta\}_m$$

$$= \sum_{m=1}^{r} [B]_m \{\delta\}_m . \tag{2.15}$$

$[B]_m$ is obtained by carrying out the appropriate partial differentiations as indicated by (2.15), and its explicit form is given in Table 2.8.

The bending and twisting moments for a plate with cylindrical ortho-tropic material are:

$$\{\sigma\} = \begin{Bmatrix} M_r \\ M_\theta \\ M_{r\theta} \end{Bmatrix} = [D][B]\{\varepsilon\} = [D]\sum_{m=1}^{r} [B]_m \{\delta\}_m \tag{2.16}$$

and the property matrix

$$[D] = \begin{bmatrix} D_r & D_1 & 0 \\ D_1 & D_\theta & 0 \\ 0 & 0 & D_k \end{bmatrix} \tag{2.17}$$

where $D_r$, $D_\theta$ are bending rigidities for directions of $r$ and $\theta$, and $D_k$ is the twisting rigidity:

$$D_r = \frac{E_r t^3}{12(1-\nu_r\nu_\theta)}, \quad D_\theta = \frac{E_\theta t^3}{12(1-\nu_r\nu_\theta)}, \quad D_k = \frac{G_{r\theta} t^3}{12},$$

$$D_1 = \nu_r D_\theta = \nu_\theta D_r .$$

TABLE 2.8. STRAIN MATRIX OF A LO2 CURVED BENDING STRIP

$$
[B]_m =
\begin{bmatrix}
\left(-\dfrac{3R}{2b'^2}+\dfrac{3}{2b'^2}\right)\Theta''_m &
\left(\dfrac{2}{b'}-\dfrac{2R}{2b'}\right)\Theta''_m &
\left(\dfrac{3R}{2b'^2}-\dfrac{3}{2b'^2}\right)\Theta''_m &
\left(\dfrac{1}{b'}-\dfrac{3R}{2b'}\right)\Theta''_m \\[2em]
-\dfrac{1}{r^2}\left(1-\dfrac{3}{4}R^2+\dfrac{1}{4}R^3\right)\Theta''_m +\dfrac{1}{r}\left(\dfrac{3R}{2b'}-\dfrac{3R^2}{4b'}\right)\Theta_m &
-\dfrac{b'}{r^2}\left(R-R^2+\dfrac{R^3}{4}\right)\Theta''_m +\dfrac{1}{r}\left(2R-1-\dfrac{3R^2}{4}\right)\Theta_m &
-\dfrac{1}{r^2}\left(\dfrac{3}{4}R^2-\dfrac{1}{4}R^3\right)\Theta''_m +\dfrac{1}{r}\left(\dfrac{3R^2}{4b'}-\dfrac{3R}{2b'}\right)\Theta_m &
-\dfrac{b'}{r^2}\left(\dfrac{R^3}{4}-\dfrac{R^2}{2}\right)\Theta''_m +\dfrac{1}{r}\left(R-\dfrac{3R^2}{4}\right)\Theta_m \\[2em]
\dfrac{2}{r}\left(\dfrac{3R}{2b'}-\dfrac{3R^2}{4b'}\right)\Theta'_m +\dfrac{2}{r^2}\left(1-\dfrac{3}{4}R^2+\dfrac{1}{4}R^3\right)\Theta'_m &
\dfrac{2}{r}\left(-1+2R-\dfrac{3R^2}{4}\right)\Theta'_m +\dfrac{2b'}{r^2}\left(R-R^2+\dfrac{R^3}{4}\right)\Theta'_m &
\dfrac{2}{r}\left(\dfrac{-3R}{2b'}+\dfrac{3R^2}{4b'}\right)\Theta'_m +\dfrac{2}{r^2}\left(\dfrac{3}{4}R^2-\dfrac{1}{4}R^3\right)\Theta'_m &
\dfrac{2}{r}\left(R-\dfrac{3R^2}{4}\right)\Theta_m +\dfrac{2b'}{r^2}\left(\dfrac{R^3}{4}-\dfrac{R^2}{2}\right)\Theta'_m
\end{bmatrix}
$$

For isotropic material, we have the familiar plate constants

$$D_r = D_\theta = D = \frac{Et^3}{12(1-v^2)}, \quad v_r = v_\theta = v, \quad D_k = \frac{1-v}{2} D.$$

All the ingredients are now ready, and it is possible to compute the stiffness and load matrices through the following equations:

$$[S]_{mn} = \int_0^\alpha \int_{r_1}^{r_2} [B]_m^T [D] [B]_n \, r \, d\theta \, dr, \tag{2.18}$$

$$\{F\}_m = \int_0^\alpha \int_{r_1}^{r_2} 2[N]_m \, qr \, d\theta \, dr. \tag{2.19}$$

TABLE 2.9. STIFFNESS MATRIX OF A LO2 CURVED BENDING STRIP

$[S]_{mn} =$

| | | | |
|---|---|---|---|
| $D_r B_{11} B_{11}$ $+ D_1 B_{21} B_{11}$ $+ D_1 B_{11} B_{21}$ $+ D_\theta B_{21} B_{21}$ $+ D_{r\theta} B_{31} B_{31}$ | $D_r B_{11} B_{12}$ $+ D_1 B_{21} B_{12}$ $+ D_1 B_{11} B_{22}$ $+ D_\theta B_{21} B_{22}$ $+ D_{r\theta} B_{31} B_{32}$ | $D_r B_{11} B_{13}$ $+ D_1 B_{21} B_{13}$ $+ D_1 B_{11} B_{23}$ $+ D_\theta B_{21} B_{23}$ $+ D_{r\theta} B_{31} B_{33}$ | $D_r B_{11} B_{14}$ $+ D_1 B_{21} B_{14}$ $+ D_1 B_{11} B_{24}$ $+ D_\theta B_{21} B_{24}$ $+ D_{r\theta} B_{31} B_{34}$ |
| $D_r B_{12} B_{11}$ $+ D_1 B_{22} B_{11}$ $+ D_1 B_{12} B_{21}$ $+ D_\theta B_{22} B_{21}$ $+ D_{r\theta} B_{32} B_{31}$ | $D_r B_{12} B_{12}$ $+ D_1 B_{22} B_{12}$ $+ D_1 B_{12} B_{22}$ $+ D_\theta B_{22} B_{22}$ $+ D_{r\theta} B_{32} B_{32}$ | $D_r B_{12} B_{13}$ $+ D_1 B_{22} B_{13}$ $+ D_1 B_{12} B_{23}$ $+ D_\theta B_{22} B_{23}$ $+ D_{r\theta} B_{32} B_{33}$ | $D_r B_{12} B_{14}$ $+ D_1 B_{22} B_{14}$ $+ D_1 B_{12} B_{24}$ $+ D_\theta B_{22} B_{24}$ $+ D_{r\theta} B_{32} B_{34}$ |
| $D_r B_{13} B_{11}$ $+ D_1 B_{23} B_{11}$ $+ D_1 B_{13} B_{21}$ $+ D_\theta B_{23} B_{21}$ $+ D_{r\theta} B_{33} B_{31}$ | $D_r B_{13} B_{12}$ $+ D_1 B_{23} B_{12}$ $+ D_1 B_{13} B_{22}$ $+ D_\theta B_{23} B_{22}$ $+ D_{r\theta} B_{33} B_{32}$ | $D_r B_{13} B_{13}$ $+ D_1 B_{23} B_{13}$ $+ D_1 B_{13} B_{23}$ $+ D_\theta B_{23} B_{23}$ $+ D_{r\theta} B_{33} B_{33}$ | $D_r B_{13} B_{14}$ $+ D_1 B_{23} B_{14}$ $+ D_1 B_{13} B_{24}$ $+ D_\theta B_{23} B_{24}$ $+ D_{r\theta} B_{33} B_{34}$ |
| $D_r B_{14} B_{11}$ $+ D_1 B_{24} B_{11}$ $+ D_1 B_{14} B_{21}$ $+ D_\theta B_{24} B_{21}$ $+ D_{r\theta} B_{34} B_{31}$ | $D_r B_{14} B_{12}$ $+ D_1 B_{24} B_{12}$ $+ D_1 B_{14} B_{22}$ $+ D_\theta B_{24} B_{22}$ $+ D_{r\theta} B_{34} B_{32}$ | $D_r B_{14} B_{13}$ $+ D_1 B_{24} B_{13}$ $+ D_1 B_{14} B_{23}$ $+ D_\theta B_{24} B_{23}$ $+ D_{r\theta} B_{34} B_{33}$ | $D_r B_{14} B_{14}$ $+ D_1 B_{24} B_{14}$ $+ D_1 B_{14} B_{24}$ $+ D_\theta B_{24} B_{24}$ $+ D_{r\theta} B_{34} B_{34}$ |

($B_{ij} B_{kl}$ refers to product of the $m$th term and $n$th term coefficients in the strain matrix.)

$$[S]_{mn} = \int_0^\alpha \int_{r_1}^{r_2} [S]_{mn} r \, dr \, d\theta.$$

Due to the presence of $1/r^n$ terms it is more expedient to integrate the above expressions numerically both with respect to $dr$ and $d\theta$. Thus the stiffness matrix listed in Table 2.9 is still in an incomplete form, and numerical integration using Gaussian quadrature or Simpson's rule remains to be performed.

## 2.4. SKEWED PLATE STRIP SIMPLY SUPPORTED AT TWO OPPOSITE SIDES

The stiffness matrix for a skewed plate strip can be developed in the same way as for a rectangular strip except that a skew coordinate $(\xi, \eta)$ system should now be used. The relationship between the skew coordinate system and the Cartesian coordinate system is given by the following transformation:

$$\begin{Bmatrix} \xi \\ \eta \end{Bmatrix} = \begin{bmatrix} 1 & -\cot \beta \\ 0 & \operatorname{cosec} \beta \end{bmatrix} \begin{Bmatrix} x \\ y \end{Bmatrix} \tag{2.20}$$

from which another transformation matrix relating the curvatures in the wo sets of coordinate systems can be derived:

$$\{\varepsilon\} = \begin{bmatrix} -\dfrac{\partial^2 w}{\partial x^2} \\[2mm] -\dfrac{\partial^2 w}{\partial y^2} \\[2mm] 2\dfrac{\partial^2 w}{\partial x\, \partial y} \end{bmatrix} = [C] \begin{bmatrix} -\dfrac{\partial^2 w}{\partial \xi^2} \\[2mm] -\dfrac{\partial^2 w}{\partial \eta^2} \\[2mm] 2\dfrac{\partial^2 w}{\partial \xi\, \partial \eta} \end{bmatrix} = [C] \{\bar{\varepsilon}\} \tag{2.21a}$$

or
$$[B]_m = [C]\,[B]_m \tag{2.21b}$$

in which the strain transformation matrix

$$[C] = \begin{bmatrix} 1 & 0 & 0 \\[2mm] \dfrac{1}{\tan^2 \beta} & \dfrac{1}{\sin^2 \beta} & \dfrac{1}{\sin \beta \, \tan \beta} \\[2mm] \dfrac{2}{\tan \beta} & 0 & \dfrac{1}{\sin \beta} \end{bmatrix}. \tag{2.21c}$$

It is not possible to use (2.6) directly to obtain the stiffness matrix of a skew strip because all the strain components in $[B]_m$ are given in Cartesian coordinates, while the displacement function is now written in skew coordinates. An intermediate step of coordinate transformation is therefore necessary. We have, for a simply supported strip,

$$[S]_{mn} = \int [B]_m^T[D] \, [B]_m \, dx \, dy$$

$$= \int [\bar{B}]_m^T[C]^T \, [D] \, [C] \, [\bar{B}]_m \sin \beta \, d\xi \, d\eta \qquad (2.21d)$$

The boundary conditions at the simply supported edge require that the deflection $w$ and the normal curvature $\partial^2 w / \partial n^2$ should both vanish. It was found that if the $x$ and $y$ terms in (2.1) and (2.9a) are simply replaced by $\xi$ and $\eta$ respectively, the condition $\partial^2 w / \partial n^2$ is not satisfied fully because we have only $w = 0$, $\partial^2 w / \partial \eta^2 = 0$, $\partial^2 w / \partial \xi^2 = 0$. However, according to Brown and Ghali,[15] satisfactory results can be obtained by using the HO2 (although not for LO2, which, however, has been used successfully for vibration problems) type of displacement function for uniformly loaded slabs with up to about 45° skew (see tables 2, 3, 4 of reference 15). For skew angles ($\beta'$) greater than 45°, the results start to deviate more strongly from the other results given by series solution, finite difference, and finite element. No example on a skew slab under concentrated load has been reported.

## 2.5. STIFFNESS MATRIX OF A BEAM IN BENDING AND TORSION

Many plate structures are stiffened with beams and very often cannot be approximated as orthotropic plates, especially when the beams are unequally spaced or are far apart from each other. A common example is a slab bridge stiffened by two edge beams.

The composite action of slab and beam can be included in a finite strip analysis in one of the following ways:

(i) A beam is considered as a one-dimensional structure along a nodal line and its stiffness assembled with the plate stiffness accordingly[4, 16]. Some calculated or estimated flexural rigidity $EI$ and torsional rigidity $GJ$ of a T-section formed by the beam and a certain width of the plate is used for computing the beam stiffnesses. This procedure is simple, although approximate, and involves no increase in the problem size.

(ii) The approach outlined in (*i*) is used, but instead of some estimated composite section the actual section of the beam (excluding its flanges) is used to compute the rigidities. The stiffness matrix is then transformed to account for the eccentricity between plate middle surface and and beam axis. Thus if the nodal displacement parameters of the beam and the slab are denoted by $\{\bar{\delta}\}$ and $\{\delta\}$ respectively, the two of them are obviously connected through some transformation involving some eccentricities, i.e.

$$\{\bar{\delta}\} = [H]\{\delta\}. \tag{2.22a}$$

The transformed stiffness matrix $S$ of the beam can then be written as

$$[S] = [H]^T[\bar{S}][H] \tag{2.22b}$$

This procedure involves no increase in the number of nodal lines but an increase in the DOF at each line since in-plane displacements and membrane forces must now be included in the formulation to cope with the shear lag phenomenon. The results obtained are quite accurate.[17-19]

(iii) The idealized structure is assumed to consist entirely of strips subjected to in-plane as well as bending forces. Such a procedure is very accurate but requires the solution of double the number of unknowns when compared with (i), and has the additional advantage of being applicable to deep beams and beams with hollow sections. A detailed description will be presented in Chapter 4.

In this section the first approach will be discussed in detail.

The stiffness matrix of a beam is formulated in the same way as that for a strip, and for compatibility, both plate strip and beam should have the same variation of deflection at the connecting nodal line. Consequently, the displacements of the beam should be written as

$$\{f\} = \begin{Bmatrix} w \\ \theta \end{Bmatrix} = \sum_{m=1}^{r} Y_m \begin{Bmatrix} w_{bm} \\ \theta_{bm} \end{Bmatrix}. \tag{2.23}$$

Going through the minimization process, the stiffness matrix can be obtained as

$$[S]_{mn} = \begin{bmatrix} k_{mn}^f & 0 \\ 0 & k_{mn}^t \end{bmatrix} \tag{2.24a}$$

in which $k_{mn}^f$ and $k_{mn}^t$ are flexural and torsional stiffness coefficients respectively given by

$$\left.\begin{array}{l} k_{mn}^f = \displaystyle\int_0^a EI Y_m'' Y_n'' \, dy, \\[2mm] k_{mn}^t = \displaystyle\int_0^a GJ Y_m' Y_n' \, dy. \end{array}\right\} \qquad (2.24b)$$

For a simply supported beam, the series decouples when $EI$ and $GJ$ are kept constant, and the stiffness coefficients become simply

$$\left.\begin{array}{l} k_{mn}^f = \dfrac{(m\pi)^4}{2} \dfrac{EI}{a^3}, \\[3mm] k_{mn}^t = \dfrac{(m\pi)^2}{2} \dfrac{GJ}{a}. \end{array}\right\} \qquad (2.24c)$$

## 2.6. NUMERICAL EXAMPLES

### 2.6.1. SIMPLY SUPPORTED ISOTROPIC SLAB UNDER UNIFORM LOAD[3] (FIG. 2.2)

The accuracy of the lower order strip LO2 is studied in this example. A series of calculations for half a slab was carried out for different mesh divisions and using different number of terms. The rate of convergence is compared with that of the finite element method in Table 2.10. It can be noticed that while convergence of deflections is extremely rapid, more strips and terms are required for the moments to converge to the exact answer. Note that the comparison of bandwidths and total number of equations will be much more disadvantageous for the finite element method if the whole slab has to be analysed, e.g. in the case of asymmetric loading.

### 2.6.2. SQUARE ISOTROPIC CLAMPED SLAB WITH ALL EDGES CLAMPED WITH A CENTRAL CONCENTRATED LOAD $P$[20]

In order to demonstrate the improved accuracy achievable by higher order strips, the slab is analysed by dividing half of it, first of all into five LO2 strips and then into only two HO2 strips. From Table 2.11 it is

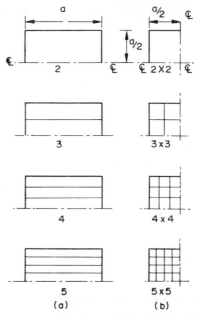

FIG. 2.2. Diagram showing mesh lines in example 2.6.1. (a) Division into
strips; (b) division into elements.

observed that the very coarse HO2 mesh actually produces results that
are slightly more accurate. In this example, all the terms of the series
are coupled together.

It should be pointed out that from the computation point of view, no
gain in efficiency has been achieved in this example through the use of
HO2 because of the 50% increase in bandwidth, which more than offsets
the small reduction in the total number of equations.

### 2.6.3. CLAMPED SEMICIRCULAR PLATE UNDER
#### UNIFORMLY DISTRIBUTED LOAD

The final example shows the use of curved strips for a rather interesting
problem. A clamped semicircular plate of radius $a$ under uniformly
distributed load is divided into eight equal strips and four terms of the
series are used in the analysis. The first strip has an inner radius equal to

TABLE 2.10. CONVERGENCE TEST USING UNIFORMLY LOADED SIMPLY SUPPORTED SLAB

$\nu = 3$

| Number of lines for slab (2 equations per line) (1) | Total number of terms (2) | Finite strip method | | | Finite element method | | |
|---|---|---|---|---|---|---|---|
| | | $w_{max}$ (3) | $M_{x\,max}$ (4) | $M_{y\,max}$ (5) | Nodal mesh for quarter slab (3 equations per node) (6) | $w_{max}$ (7) | $M_{max}$ (8) |
| 2 | 1 | 0.00414 | 0.0561 | 0.0540 | 2×2 | 0.00350 | 0.0669 |
| | 2 | 0.00409 | 0.0546 | 0.0494 | | | |
| | 3 | 0.00409 | 0.0549 | 0.0504 | | | |
| | 4 | 0.00409 | 0.0549 | 0.0504 | | | |
| 3 | 1 | 0.00411 | 0.0502 | 0.0520 | 3×3 | 0.00395 | 0.0502 |
| | 2 | 0.00406 | 0.0487 | 0.0474 | | | |
| | 3 | 0.00406 | 0.0490 | 0.0485 | | | |
| | 4 | 0.00406 | 0.0489 | 0.0481 | | | |
| 4 | 1 | 0.00411 | 0.0496 | 0.0518 | 4×4 | 0.00401 | 0.0483 |
| | 2 | 0.00406 | 0.0480 | 0.0472 | | | |
| | 3 | 0.00406 | 0.0484 | 0.0483 | | | |
| | 4 | 0.00406 | 0.0481 | 0.0479 | | | |
| 5 | 1 | 0.00411 | 0.0494 | 0.0517 | 5×5 | 0.00403 | |
| | 2 | 0.00406 | 0.0479 | 0.0472 | | | |
| | 3 | 0.00406 | 0.0482 | 0.0482 | | | |
| | 4 | 0.00406 | 0.0481 | 0.0478 | | | |
| Exact (1) | | 0.00406 | 0.0479 | 0.0479 | | 0.00406 | 0.0479 |
| Multiplier | | $qa^4/D$ | $qa^2$ | $qa^2$ | | $qa^4/D$ | $qa^2$ |

TABLE 2.11. STATIC ANALYSIS OF SLAB WITH FOUR CLAMPED SIDES

| $v = 0.3$ | Central concentrated load, $P$ | | | | | |
|---|---|---|---|---|---|---|
| | $W_{max}$ | | $-M_{x\,max}$ | | $-M_{y\,max}$ | |
| | HO2 | LO2 | HO2 | LO2 | HO2 | LO2 |
| $m = 1$ | 0.00511 | 0.00510 | −0.11838 | −0.11562 | −0.14385 | −0.14381 |
| $m = 3$ | 0.00030 | 0.00031 | −0.00949 | −0.00634 | 0.05122 | 0.05242 |
| $m = 5$ | 0.00011 | 0.00011 | +0.00331 | 0.00134 | −0.04690 | −0.04421 |
| $m = 7$ | 0.00003 | 0.00003 | −0.00198 | −0.00061 | 0.02677 | 0.02638 |
| $\Sigma$ | 0.00555 | 0.00555 | −0.12654 | −0.12129 | −0.11276 | −0.10922 |
| Exact | 0.00560 | 0.00560 | −0.12570 | −0.12570 | −0.12570 | −0.12570 |
| Multiplier | $\dfrac{Pa^2}{D}$ | | $P$ | | | |

zero and also its two radial edges forming part of a straight line; neverthe-less, no numerical dificulty is experienced because no singularity can occur when Gaussian quadrature is used in the numerical integration of the strip stiffness. This is due to the fact that end points are never used as integration points in such a scheme.

The finite strip results give a maximum centre deflection of $0.002021$ $qa^2/D$ at $r = 0.4859a$ and a maximum negative bending moment of $M_r = -0.0697\,qa^2$ at $r = a$, while the corresponding values from an ana-lytical solution by Woinowsky-Krieger[21] are found to be $0.002022qa^4/D$ and $-0.0731qa^2$. Note that the use of equal width strips is merely for convenience, and better results are usually achievable for the same number of strips if strips of unequal widths are used so that regions of steep stress gradient are served by finer meshes.

## 2.7. APPLICATION TO SLAB BRIDGES

### 2.7.1. INTRODUCTION

The analysis of slab bridges has been the centre of interest for many researchers, and many different methods of various degree of sophisti-cation have been proposed. However, most methods (apart from the finite element method, finite difference method, and grillage analysis, which all require a great deal of computational efforts) are unable to cover all the essential features of the majority of slab bridges which are commonly used in practice. Such features are listed as follows:

(i) The bridge may be straight or curved in plan, and is of constant width along the span.

(ii) The bridge is simply supported at the two ends, but may also be elastically supported by longitudinal beams and intermediate columns which need not be evenly spaced.

(iii) Transverse variations of thicknesses are not uncommon.

(iv) The structures are usually either plain isotropic slabs or are made up of a system of slabs and multiple stiffening girders. These multiple girders are usually arranged orthogonally and thus such straight or curved bridges can be treated as equivalent orthotropic or cylindrical orthotropic plates.

A popular method of analysis in Europe for right bridge decks is the orthotropic plate theory originated by Guyon[22] and Massonet,[23] and design charts have been prepared by various workers[24, 25] using only the first term of the harmonic series. A new set of design charts using fifteen terms of the series have been prepared by Cusens and Pama.[26] However, the method is less versatile because the deck is assumed to be uniform throughout, and the use of design charts, apart from having some built-in approximations, will not allow a designer to go up to any accuracy that he might judge to be adequate.

The finite element method is, of course, the most versatile technique available, and a BAPS finite element program package[27] using a triangular element developed by Bazeley et al[28] has been made available by the British Ministry of Transport. Nevertheless, the program can be shown to be relatively inefficient in analysing straight and curved bridge decks with simply supported ends. This fact will be borne out by an example later.

The finite strip method is an ideal tool for analysing bridges with the previously mentioned features. Since curved strips and skewed strips have also been developed in addition to the right strips, it is perfectly simple to incorporate all the different stiffness matrices as subroutines and to analyse right bridge, curved bridge, and skewed bridge by the same computer program. Because the terms of the series are uncoupled, a solution of any required accuracy can be achieved by simply adding in the results of the higher terms without having to forgo the results that had already been computed.

The analysis of right slab bridges by the finite strip method was first attempted by Cheung.[6] Later the method was extended to slab–beam bridges by Powell and Ogden[4] and by Cheung et al.[16] Further analysis of right slab bridges based on higher order strips were presented by Loo and Cusens.[7, 8] Finally, a curved strip was applied by Cheung[12] to the solution of curved bridge decks.

## 2.7.2. ANALYSIS OF COLUMN SUPPORTED BRIDGE DECKS (FIG. 2.3)

For bridge decks with arbitrarily spaced columns the force method or flexibility method is normally used. The released structure, which in the present case amounts to a bridge simply supported at its two ends only, is analysed under the external loading and also under the redundant forces which are either unit point loads[16] or uniformly distributed rectangular patch loads[29] at the column locations, using the standard finite strip procedure and with as many right-hand side loading vectors as there are external load cases and redundants. This procedure produces a set of displacements $\Delta$ at the column locations for all the external loadings, and also another set of displacements, or flexibility coefficients $f$ due to the unit reaction forces. In matrix notation,

$$[\Delta] = \begin{bmatrix} \Delta_{11} & \Delta_{12} & \dots & \Delta_{1n} \\ \Delta_{21} & \Delta_{22} & \dots & \Delta_{2n} \\ \dots & \dots & \dots & \dots \\ \Delta_{n1} & \Delta_{n2} & \dots & \Delta_{nn} \end{bmatrix} \qquad (2.25a)$$

$$[f] = \begin{bmatrix} f_{11} & f_{12} & \dots & f_{1n} \\ f_{21} & f_{22} & \dots & f_{2n} \\ \dots & \dots & \dots & \dots \\ f_{n1} & f_{n2} & \dots & f_{nn} \end{bmatrix} \qquad (2.25b)$$

in which $\Delta_{ij}$ is the deflection at point $i$ due to load vector $j$, and $f_{ij}$ is the deflection at point $i$ due to unit reaction force at $j$. It is possible also to include the effects due to elastic deformations of the columns and foundation settlements by modifying (2.25b) to

$$[f] = \begin{bmatrix} (f_{11}+f'_{11}) & f_{12} & \dots & f_{1n} \\ f_{21} & (f_{22}+f'_{22}) & \dots & f_{2n} \\ \dots & \dots & \dots & \dots \\ f_{n1} & f_{n2} & \dots & (f_{nn}+f'_{nn}) \end{bmatrix} \qquad (2.26)$$

where $f'_{ii}$ is the total deformation at the top of column $i$ due to column shortening and foundation settlement.

5

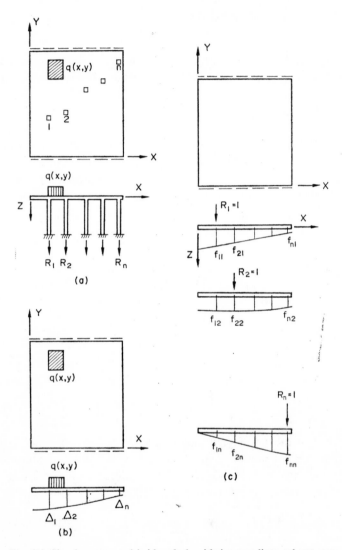

FIG. 2.3. Simply supported bridge deck with intermediate column supports and the released structure. (a) Loading and column reactions. (b) Released structure with loading. (c) Released structure under individual unit column reactions.

If the compatibility requirement of displacements at the column positions are now considered, it is obvious that we should have

$$[f][R]+[\Delta] = [0] \qquad (2.27)$$

from which the true column reactions $[R]$ can be calculated.

The final displacements and internal forces can be evaluated by multiplying and combining with the results that were obtained in the first part of the analysis.

For plates and bridges over continuous line supports, it is also possible to treat each individual span as simply supported, and continuity of slopes over the supports (at the nodal lines only) are subsequently restored. In this case the support moments are chosen as the redundant forces.[30]

### 2.7.3. SOME BRIDGE PROBLEMS

(i) A square constant thickness slab bridge[20] is analysed for a central concentrated load, using 8 LO2, 4 HO2 strips, and $4\times8$ rectangular finite elements respectively for half the bridge. The distribution coefficients

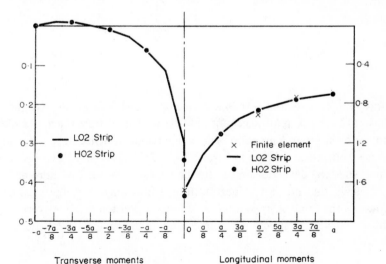

Transverse moments　　　　Longitudinal moments

FIG. 2.4. Distribution coefficients for transverse and longitudinal moments for a unit load at centre of a square bridge.

(Fig. 2.4) computed for longitudinal and transverse moments are practically identical for the three cases.

(ii) An isotropic bridge deck[11] with three rigid intermediate columns and acted upon by partially distributed loads is shown in Fig. 2.5. The bridge is analysed by a HO3 strip program and a finite element program FESS, the predecessor of the BAPS program, using eight equal strips

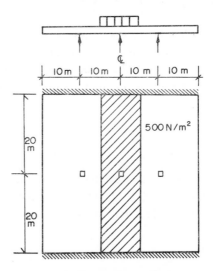

FIG. 2.5. Simply supported bridge deck with three intermediate columns.

and $8\times8$, $16\times16$, and $18\times24$ finite element meshes (Fig. 2.6). Symmetry has not been taken into account in order to demonstrate the application to the most general situation. The comparisons of deflections and transverse moments are shown in Figs. 2.7 and 2.8 respectively. It can be seen that while the deflections given by the various analyses are more or less the same, the moment values given by FESS only converge gradually towards the finite strip solution as the mesh division becomes very fine. In all fairness it should be pointed out that average nodal moment[†]

---

† The centroidal moments of all triangles connected to a particular node are summed and an average nodal value obtained by dividing the sum with the total number of triangles.

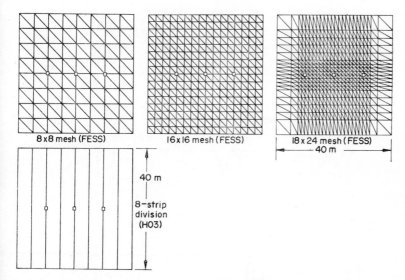

FIG. 2.6. Mesh patterns used in continuous bridge analysis (Loo[11]).
(Original unit in FPS.)

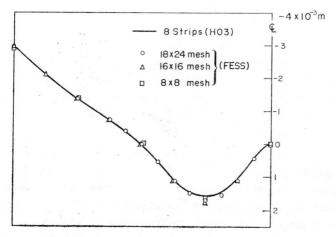

FIG. 2.7. Deflection profiles at line of columns (Loo[11]). ( Original
unit in FPS.)

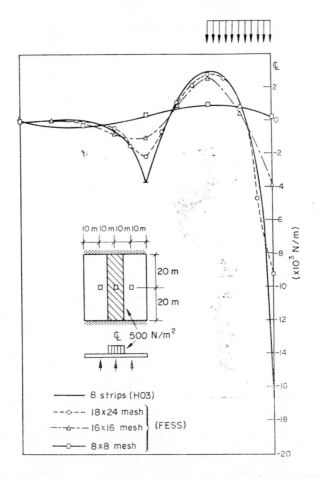

Fig. 2.8. Transverse moment profiles at line of columns (Loo[11]).
(Original unit in FPS.)

values are used in the present finite element plots, and this technique
will usually produce unfavourable results near free edges, lines of
symmetry, and edges of zones with rapid stress gradient.[31] In this example
a much better comparison could be arrived at by plotting the centroidal
values of the elements near the columns and extrapolating the results.
Based on the above discussion it is possible to say that the 16×24 mesh
results are in close agreement with those of the HO3 strips. A comparison

TABLE 2.12. COMPUTER TIME USED BY HO3 AND FESS FOR THE ANALYSIS
OF AN ISOTROPIC SQUARE PLATE IN AN ICL 4130 COMPUTER USING T3OC
SYSTEM (Loo[11])

| Details | HO3 | | FESS | | | | | |
|---|---|---|---|---|---|---|---|---|
| | 8 – strips | | 8×8 mesh | | 16×16 mesh | | 18×24 mesh | |
| | min | sec | min | sec | min | sec | min | sec |
| Data checking (FECK) | — | — | — | 10 | — | 12 | — | 20 |
| Data generating (JPUT) | — | — | 1 | 43 | 5 | 38 | 9 | 34 |
| Analysis | — | 48[a] | 5 | 6 | 25 | 26 | 47 | 09 |
| Total | — | 48 | 6 | 59 | 31 | 16 | 57 | 3 |

[a] Two load cases and fifteen harmonics (fifteen terms).

of efficiency for the various analyses is given in Table 2.12, and the results are strongly in favour of the finite strip method. All the analyses were performed on the ICL 4130 computer using T30C system, and thus the uncertain factor of relative machine speed has been eliminated.

(iii) A model bridge[16] with two beams and variable thickness cross-section is shown in Fig. 2.9a. A central concentrated load of 556 N is applied to the model and experimental results are compared with those of a LO2 finite strip analysis, in which half of the bridge is idealized by eight strips, and the flexural and torsional rigidities are assumed to be equal to that of a T-section with the flange equal to half the width of the bridge. Good agreement is observed for all comparisons made in the longitudinal direction for plate moments as well as beam stresses (Fig. 2.9b, c). Reasonable agreement is obtained for the curvature comparisons along the transverse centre line (Fig. 2.9d), and indicates that the rotational stiffness of the beam might have been overestimated.

(iv) The final example[12] deals with the analysis of uniform thickness isotropic curved bridges and the theoretical and experimental results obtained by Coull and Das[13] are used as a check on the accuracy of the finite strip method. A model bridge slab made of asbestos-cement and

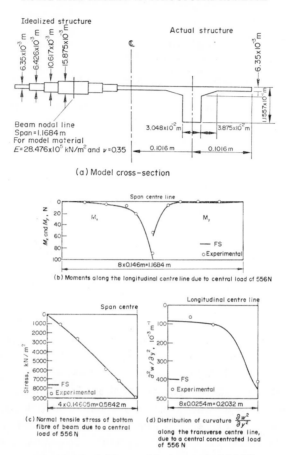

FIG. 2.9. The bridge model of example 2.7.3(iii).

with an included angle of 60° is loaded with central point loads at three different radial positions—the inner and outer edges, and at mid-radius respectively. A comparison of the deflections and bending moments along the central section is given in Fig. 2.10A and 2.10B and, in general, there is good agreement between all sets of results. The finite strip method consistently gives higher moments under the load points when sufficient number of terms are used (eight terms in this case). However, if only three terms are used, the maximum values then are much closer to the theoretical results of Coull and Das, who also took only three terms for their analysis. A point of interest is that the subsequent higher terms

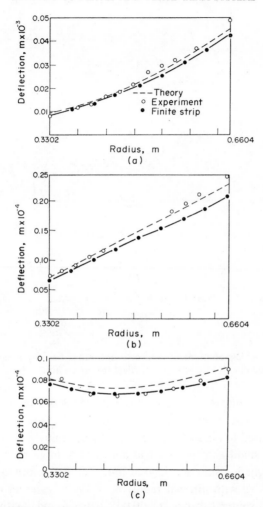

FIG. 2.10A. Radial distributions of mid-span deflections due to unit load at (a) outer edge, (b) mid-radius, (c) inner edge; asbestos-cement model

(e.g. after the third or fourth term) only affect results in the near vicinity of the point load. From the deflection diagrams it can be concluded that for a curved bridge the outer edge is much more flexible than the inner edge.

A typical run for one loading, using eight terms and eight strips, took

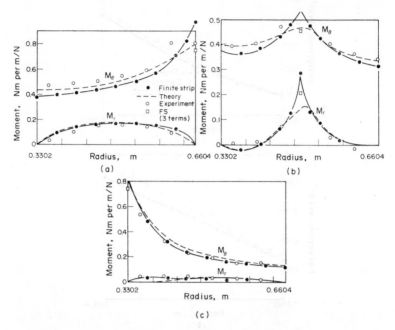

FIG. 2.10B. Distributions of tangential and radial bending moments at
mid-span due to load at (a) outer edge, (b) mid-radius, (c) inner edge;
asbestos-cement model.

under half a minute on the medium speed IBM 360–50 computer.

It is also of considerable interest to note that the curved strip program can also be used to analyse right bridges by assigning a "curved" bridge with a very small subtended angle and a very large radius. In a test example in which a square bridge of unit span and width is analysed as a curved bridge with internal radius $R_i = 199.5$, external radius $R_e = 200.5$, and subtended angle $\alpha = 0.005$, it was found that the maximum error in deflection is 0.7%, while no difference can be detected for all comparisons made at mid-radius. No numerical instability has ever been experienced for other examples in which even larger radius values have been used.

### REFERENCES

1. S. P. Timoshenko and S. Woinowsky-Krieger, *Theory of Plates and Shells*, 2nd edn., McGraw-Hill, 1959.
2. Y. K. Cheung, Finite strip method in the analysis of elastic plates with two opposite ends, *Proc. Instn. Civ. Engrs* **40**, 1–7 (May 1968).

3. Y. K. CHEUNG, Finite strip method analysis of elastic slabs, *Am. Soc. Civ. Engrs* **94**, EM6, 1365–78 (December 1968).

4. G. H. POWELL and D. W. OGDEN, Analysis of orthotropic bridge decks, *Am. Soc. Civ. Engrs* **95**, ST5, 909–23 (May 1969).

5. J. G. ERGATOURDIS, B. M. IRONS, and O. C. ZIENKIEWICZ, Curved, isoparametric, quadrilateral elements for finite element analysis, *Int. J. Solids Struct.* **4**, 31–42 (1968).

6. Y. K. CHEUNG, *Orthotropic Right Bridges by the Finite Strip Method*, Concrete Bridge Design, ACI Publications SP-26, 182–205, 1971.

7. Y. C. LOO and A. R. CUSENS, A refined finite strip method for the analysis of orthotropic plates, *Proc. Instn. Civ. Engrs* **40**, 85–91 (January 1971).

8. Y. C. LOO and A. R. CUSENS, Developments of the finite strip method in the analysis of bridge decks, *Developments in Bridge Design and Construction* (ed. Rockey *et al.*), Crosby Lockwood, 1971.

9. Y. K. CHEUNG and M. S. CHEUNG, Flexural vibrations of rectangular and other polygonal plates, *Am. Soc. Civ. Engrs* **97**, EM 2, 391–411 (April 1971).

10. M. S. CHEUNG, Finite strip analysis of structures, PhD thesis, Department of Civil Engineering, University of Calgary, 1971.

11. Y. C. LOO, Developments and applications of the finite strip method in the analysis of right bridge decks, PhD thesis, Department of Civil Engineering, University of Dundee, 1972.

12. Y. K. CHEUNG, The analysis of curvilinear orthotropic curved bridge decks, *Publications, International Association for Bridges and Structural Engineering* 29-II, 1969.

13. A. COULL and P. C. DAS, Analysis of curved bridge decks, *Proc. Inst. Civ. Engrs* **37**, 75–85 (May 1967).

14. T. H. YONEZAWA, Moment and force vibrations in curved girder bridges, *Am. Soc. Civ. Engrs* **88**, EM1, 1–21 (February 1962).

15. T. G. BROWN and A. GHALI, Finite strip analysis of skew plates, *Proc. McGill – EIC Specialty Conference on the Finite Element Method in Civil Engineering, Montreal, June 1972.*

16. M. S. CHEUNG, Y. K. CHEUNG, and A. GHALI, Analysis of slab and girder bridges by the finite strip method, *Building Sci.* **5**, 95–104 (1970).

17. W. C. GASTAFSON and R. N. WRIGHT, Analysis of skewed composite girder bridges, *Am. Soc. Civ. Engrs* **94**, ST4, 919–41 (April 1968).

18. J. D. DAVIES, I. J. SOMERVAILLE, and O. C. ZIENKIEWICZ, Analysis of various types of bridges by finite element method, *Developments in Bridge Design and Construction* (ed. Rockey *et al.*), Crosby Lockwood, 1971.

19. R. G. SISODIYA, A. GHALI, and Y. K. CHEUNG, Diaphragms in single and double cell box girder bridges with varying angle of skew, *ACI Jl* 415–19 (July 1972).

20. M. S. CHEUNG and Y. K. CHEUNG, Static and dynamic behaviour of rectangular strips using higher order finite strips, *Building Sci.* **7** (3) 151–8 (September 1972).

21. S. WOINOWSKY-KRIEGER, Clamped semicircular plate under uniform bending load, *J. appl. Mech. Trans. Am. Soc. mech. Engrs* **22** (1955).

22. Y. GUYON, Calcul des ponts larges à poutres multiples solidarisées par les entretoises, *Annuales des Ponts et Chausses* **24** (5) (September–October 1946).

23. C. MASSONET, Method of calculation of bridge with several longitudinal beams taking into account their torsional resistance, *Publications, International Association for Bridges and Structural Engineering*, 10, 1950.

24. P. B. MORICE and G. LITTLE, *Analysis of Right Bridge Decks Subjected to Abnormal Loading*, Cement and Concrete Association, London, 1956.
25. R. E. ROWE, *Concrete Bridge Design*, C. R. Books Ltd., London, 1962.
26. A. R. CUSENS and R. P. PAMA, *Design Curves for the Approximate Determination of Bending Moments in Orthotropic Bridge Decks*, Civil Engineering Department, Univ. of Dundee, March 1970.
27. K. SRISKANDAN, Case studies in the use of the BAPS finite element program package, in *Developments in Bridge Design and Construction* (ed. Rockey *et al.*), Crosby Lockwood, 1971.
28. G. P. BAZELEY, Y. K. CHEUNG, B. M. IRONS, and O. C. ZIENKIEWICZ, Triangular elements in plate bending—conformin gand non-conforming solutions, *Proceedings of the Conference on Matrix Methods in Structural Mechanics (Wright-Patterson Air Force Base, Ohio, 1965)*, 1967.
29. R. P. PAMA and A. R. CUSENS, *A Load Distribution Method for Analysing Statically Indeterminate Concrete Bridge Decks*, Concrete Bridge Design, ACI Publications SP-26, 599–633, 1971.
30. A. GHALI and G. S. TADROS, *On Finite Strip Analysis of Continuous Plates*, Research Report No. CE-72-11, Department of Civil. Engineering, University of Calgary, March 1972.
31. Y. K. CHEUNG, I. P. KING, and O. C. ZIENKIEWICZ, Slab bridges with arbitrary shape and boundary conditions: a general method of analysis based on finite elements, *Proc. Instn. Civ. Engrs* **40**, 9–36 (May 1968).

# CHAPTER 3

# *Plane stress analysis*

## 3.1. INTRODUCTION

In this chapter solutions for plane stress problems in two-dimensional elasticity will be presented, with plane strain problems similarly tackled by a suitable conversion of elastic constants.[†] The scope of direct application for such plane stress strips is limited but may include deep beams, layered or sandwich beams, layered foundation, etc. Nevertheless, they are of considerable importance when used in conjunction with bending strips in which combined flat shell strips can be formed for the analysis of straight or curved stiffened plates, folded plate structures, and box girder bridges.

Various types of strips with two simply supported or two clamped ends have been formulated and each of them will be discussed in detail in subsequent paragraphs. The simply supported strip is by far the most important and useful one, and, similar to the simply supported bending strip, the series used in the displacement functions also decouple, and therefore each term of the series can be analysed separately.

In Section 1.3.1 it has been mentioned that for the series part of the displacement field both $Y_m$ and $Y'_m$ are present in the analysis of two-dimensional elasticity problems, where, in general, $Y_m$ is used for $u$ and $Y'_m$ for $v$ displacements (Fig. 3.1). The above formulation is based on the observation of the relationship commonly used in the small deflection theory of beams, in which the transverse deflection $u$ is related to the longitudinal displacement $v$ through

$$v = A \frac{du}{dy}, \tag{3.1}$$

[†] A plane stress stiffness matrix can be used to analyse plane strain problems by assigning the Young modulus as $E/(1-v^2)$ and the Poisson ratio as $v/(1-v)$.

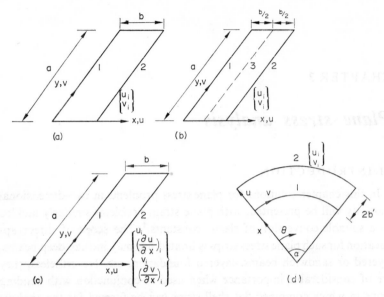

FIG. 3.1. (a) LO2 strip. (b) HO3 strip. (c) HO2 strip. (d) Curved LO2 strip.

and therefore the general form for the displacements is given as

$$
\left.
\begin{aligned}
u &= \sum_{m=1}^{r} f_m^u(x)Y_m, \\
v &= \sum_{m=1}^{r} f_m^v(x) \frac{a}{\mu_m} Y_m'.
\end{aligned}
\right\}
\tag{3.2}
$$

The above formulation can also be interpreted from the displacement functions for the simply supported case in which a sine series is used for the $u$ displacement and a cosine series for the $v$ displacement.

For vibration problems, $Y_m$ can take up any one of the six functions (with the appropriate $\mu_m$ values) listed in Section 1.3.1, and accurate predictions of frequencies and mode shapes have been obtained for all cases. In static analysis, however, (3.2) has been found to be valid for the simply supported case only.

For example, the function for the clamped–clamped case requires that $Y_m(0) = Y_m'(0) = Y_m(a) \, Y_m'(a) = 0$, and this means that the shear stress which is given by

$$\tau_{xy} = G\left(\frac{\partial u}{\partial y} + \frac{\partial v}{\partial x}\right)$$

$$= G\left(\sum_{m=1}^{r} f_m^u(x)Y_m' + \sum_{m=1}^{r} \frac{\partial[f_m^v(x)]}{\partial x}\frac{a}{\mu_m}Y_m'\right) \tag{3.3}$$

will be equal to zero at the two ends. This is obviously an impossible situation in a beam problem because it means that no reaction will be provided to resist the vertical load at the supports.

From the above discussion it can be seen that for static analysis the assumed function should satisfy the natural boundary conditions[1] as well as the displacement boundary conditions. An alternative set of functions for the clamped–clamped case was developed and applied successfully to several problems.

The two sets of displacement functions which can be used in static analysis are listed below.

(i) Both ends simply supported $\left[u = 0, \ \sigma_y = E_y\left(\frac{\partial v}{\partial y} + v_x\frac{\partial u}{\partial x}\right) = 0 \text{ at}\right.$

$y = 0$ and $y = a\Big]$.

$$\left.\begin{aligned} u &= \sum_{m=1}^{r} f_m^u(x)\,Y_m = \sum_{m=1}^{r} f_m^u(x)\sin k_m y, \\ v &= \sum_{m=1}^{r} f_m^v(x)\,\frac{a}{\mu_m}Y_m' = \sum_{m=1}^{r} f_m^v(x)\cos k_m y. \end{aligned}\right\} \tag{3.4}$$

(ii) Both ends clamped ($u = v = 0$ at $y = 0$ and $y = a$).

$$\left.\begin{aligned} u &= \sum_{m=1}^{r} f_m^u(x)\sin k_m y, \\ v &= \sum_{m=1}^{r} f_m^v(x)\sin k_{m+1} y, \end{aligned}\right\} \tag{3.5}$$

in which $k_{m+1}$ refers to $[(m+1)\pi]/a$.

## 3.2. RECTANGULAR PLANE STRESS STRIP

The first rectangular plane stress strip was developed by Cheung[2] using a linear polynomial of $x$ (1.9a) for both the $u$ and $v$ displacements. This lower order strip with constant strain $\varepsilon_x$ in the transverse direction tend to approximate the true $\sigma_x$ and $\varepsilon_x$ curve in a stepwise fashion if the results are output at the nodal lines and plotted accordingly. In order to avoid such abrupt stress and strain jumps at the interface of the strips, all $\sigma_x$ and $\varepsilon_x$ values should only be plotted at the centre line ($x = \frac{1}{2}b$) of the strip.

A higher order strip having one additional internal nodal line was later presented by Loo and Cusens.[3] In this strip a parabolic variation of displacements across the section is assumed and it is found, as expected, that stress jumps at the interface of adjoining strips are much smaller than those found in the lower order strip analysis. In order to facilitate computation and to reduce the bandwidth of the overall matrix, the parameters associated with the internal nodal line are usually eliminated through static condensation before the assembly stage.

In a recent paper by Yoshida and Oka[4] several plane stress strips using $(u, \partial u/\partial x, v)$, $(u, v, \partial v/\partial x)$ and $(u, \partial u/\partial x, v, \partial v/\partial x)$ as nodal parameters respectively were suggested. After making some numerical comparisons it was concluded that the displacement functions with $(u, \partial u/\partial x, v, \partial v/\partial x)$ as nodal parameters should be adopted because of its superior accuracy. It should be pointed out, however, that such a strip cannot be used directly for problems in which an abrupt change in plate thickness or property occurs, simply because at such locations the strains are always discontinuous.

The higher order strips have been formulated for the simply supported case only. Although no difficulty should be encountered in the extension to cover other boundary conditions, the materials presented in subsequent sections on higher order strip will nevertheless be limited to that of a simply supported strip only.

### 3.2.1. LOWER ORDER STRIP WITH TWO NODAL LINES (LO2)

#### (a) Displacement functions

A typical strip is shown in Fig. 3.1a in which only $u$ and $v$ displacement are used as nodal parameters. A suitable shape function is therefore found in (1.9a), and it is possible to write

$$
\left.
\begin{aligned}
u &= \sum_{m=1}^{r} [(1-\bar{x})\,(\bar{x})] \begin{Bmatrix} u_1 \\ u_2 \end{Bmatrix}_m Y_m, \\[2ex]
v &= \sum_{m=1}^{r} [(1-\bar{x})\,(\bar{x})] \begin{Bmatrix} v_1 \\ v_2 \end{Bmatrix}_m \frac{a}{\mu_m} Y'_m,
\end{aligned}
\right\}
\tag{3.6}
$$

in which $\bar{x} = x/b$.

The above equation can be written in matrix form as

$$
\{f\} = \begin{Bmatrix} u \\ v \end{Bmatrix} = \sum_{m=1}^{r}
\begin{bmatrix}
(1-\bar{x})Y_m & 0 & (\bar{x})Y_m & 0 \\[2ex]
0 & (1-\bar{x})\dfrac{a}{\mu_m}Y'_m & 0 & (\bar{x})\dfrac{a}{\mu_m}Y'_m
\end{bmatrix}
\begin{Bmatrix} u_1 \\ v_1 \\ u_2 \\ v_2 \end{Bmatrix}_m
$$

$$
= \sum_{m=1}^{r} [N]_m \{\delta\}_m.
\tag{3.7}
$$

Note that the constant $a/\mu_m$ before $Y'_m$ is included simply for convenience because once the differentiation for $Y_m$ is actually carried out, this quantity would be cancelled out.

#### (b) Strains

The strains for a plane stress problem are the two direct strains and a shear strain, and are given as

$$
\{\varepsilon\} = \begin{Bmatrix} \varepsilon_x \\ \varepsilon_y \\ \gamma_{xy} \end{Bmatrix} =
\begin{Bmatrix}
\dfrac{\partial u}{\partial x} \\[2ex]
\dfrac{\partial v}{\partial y} \\[2ex]
\dfrac{\partial u}{\partial y} + \dfrac{\partial v}{\partial x}
\end{Bmatrix}
= \sum_{m=1}^{r} [B]_m \{\delta\}_m
\tag{3.8}
$$

The strain matrix $[B]_m$ is obtained by performing the appropriate differentiation on the displacements in (3.7), and its explicit form is as follows:

$$[B]_m = \begin{bmatrix} \dfrac{-1}{b} Y_m & 0 & \dfrac{1}{b} Y_m & 0 \\[2ex] 0 & (1-\bar{x})\dfrac{a}{\mu_m} Y_m'' & 0 & (\bar{x})\dfrac{a}{\mu_m} Y_m'' \\[2ex] (1-\bar{x})Y_m' & -\dfrac{1}{b}\dfrac{a}{\mu_m} Y_m' & (\bar{x})Y_m' & \dfrac{1}{b}\dfrac{a}{\mu_m} Y_m' \end{bmatrix}. \quad (3.9)$$

(c) Stresses

The stresses corresponding to $\varepsilon_x$, $\varepsilon_y$, and $\gamma_{xy}$ for orthotropic materials are $\sigma_x$, $\sigma_y$, and $\tau_{xy}$, and the two sets are connected to each other through the following relationship:

$$\{\sigma\} = [D]\{\varepsilon\}$$

$$= [D]\sum_{m=1}^{r} [B]_m \{\delta\}_m \quad (3.10a)$$

in which the property matrix for orthotropic materials is

$$[D] = \begin{bmatrix} \dfrac{E_x}{1-\nu_x\nu_y} & \dfrac{\nu_x E_y}{1-\nu_x\nu_y} & 0 \\[2ex] \dfrac{\nu_x E_y}{1-\nu_x\nu_y} & \dfrac{E_y}{1-\nu_x\nu_y} & 0 \\[2ex] 0 & 0 & G_{xy} \end{bmatrix}. \quad (3.10b)$$

The conversion of (3.10b) to the isotropic case has been discussed previously in Section 2.2 and will not be repeated here.

(d) Stiffness matrix and consistent load matrix

Since the thickness of a strip is assumed to be constant, (1.30)—which is used for the computation of all stiffness matrices—can be simplified into the following form:

$$[S]_{mn} = t \int [B]_m^T [D] [B]_n \, d\,(\text{area}). \quad (3.11)$$

TABLE 3.1. IN-PLANE STIFFNESS MATRIX OF A RECTANGULAR STRIP

$$[S]_{mn} = t \begin{bmatrix}
\begin{array}{l} +K_2\left(\dfrac{-1}{2C_1}\right)I_5 \\ +K_4\left(\dfrac{1}{2C_1}\right)I_2 \end{array} &
\begin{array}{l} +K_1\left(\dfrac{-1}{b}\right)I_1 \\ +K_4\left(\dfrac{b}{6}\right)I_2 \end{array} &
\begin{array}{l} +K_2\left(\dfrac{-1}{2C_1}\right)I_5 \\ +K_4\left(\dfrac{-1}{2C_1}\right)I_2 \end{array} &
\begin{array}{l} +K_1\left(\dfrac{1}{b}\right)I_1 \\ +K_4\left(\dfrac{b}{3}\right)I_2 \end{array} \\[4ex]

\begin{array}{l} +K_3\left(\dfrac{b}{6C_1C_2}\right)I_4 \\ +K_4\left(\dfrac{-1}{bC_1C_2}\right)I_2 \end{array} &
\begin{array}{l} +K_2\left(\dfrac{1}{2C_2}\right)I_3 \\ +K_4\left(\dfrac{-1}{2C_2}\right)I_2 \end{array} &
\begin{array}{l} +K_3\left(\dfrac{b}{3C_1C_2}\right)I_4 \\ +K_4\left(\dfrac{1}{bC_1C_2}\right)I_2 \end{array} &
\begin{array}{l} +K_2\left(\dfrac{-1}{2C_2}\right)I_3 \\ +K_4\left(\dfrac{-1}{2C_2}\right)I_2 \end{array} \\[4ex]

\begin{array}{l} +K_2\left(\dfrac{1}{2C_1}\right)I_5 \\ +K_4\left(\dfrac{1}{2C_1}\right)I_2 \end{array} &
\begin{array}{l} +K_1\left(\dfrac{1}{b}\right)I_1 \\ +K_4\left(\dfrac{b}{3}\right)I_2 \end{array} &
\begin{array}{l} +K_2\left(\dfrac{1}{2C_1}\right)I_5 \\ +K_4\left(\dfrac{-1}{2C_1}\right)I_2 \end{array} &
\begin{array}{l} +K_1\left(\dfrac{-1}{b}\right)I_1 \\ +K_4\left(\dfrac{b}{6}\right)I_2 \end{array} \\[4ex]

\begin{array}{l} +K_3\left(\dfrac{b}{3C_1C_2}\right)I_4 \\ +K_4\left(\dfrac{1}{bC_1C_2}\right)I_2 \end{array} &
\begin{array}{l} +K_2\left(\dfrac{1}{2C_2}\right)I_3 \\ +K_4\left(\dfrac{1}{2C_2}\right)I_2 \end{array} &
\begin{array}{l} +K_3\left(\dfrac{b}{6C_1C_2}\right)I_4 \\ +K_4\left(\dfrac{-1}{bC_1C_2}\right)I_2 \end{array} &
\begin{array}{l} +K_2\left(\dfrac{-1}{2C_2}\right)I_3 \\ +K_4\left(\dfrac{1}{2C_2}\right)I_2 \end{array}
\end{bmatrix}$$

$$I_1 = \int_0^a Y_m Y_n \, dy, \quad I_2 = \int_0^a Y'_m Y'_n \, dy, \quad I_3 = \int_0^a Y_m Y''_n \, dy, \quad I_4 = \int_0^a Y''_m Y'_n \, dy, \quad I_5 = \int_0^a Y''_m Y''_n \, dy,$$

$$K_1 = \frac{E_x}{1-\nu_x\nu_y}, \quad K_2 = \frac{\nu_x E_y}{1-\nu_x\nu_y}, \quad K_3 = \frac{E_y}{1-\nu_x\nu_y}, \quad K_4 = G_{xy}, \quad C_1 = \frac{\mu_m}{a}, \quad C_2 = \frac{\mu_n}{a}$$

TABLE 3.2. IN-PLANE STIFFNESS MATRIX OF SIMPLY SUPPORTED LO2 RECTANGULAR STRIP

$$[S]_{mm} = t \begin{bmatrix} \dfrac{aE_1}{2b} + \dfrac{abk_m^2 G}{6} & & & \text{Symmetrical} \\[2ex] \dfrac{ak_m\nu_x E_2}{4} - \dfrac{ak_m G}{4} & \dfrac{abk_m^2 E_2}{6} + \dfrac{aG}{2b} & & \\[2ex] -\dfrac{aE_1}{2b} + \dfrac{abk_m^2 G}{12} & -\dfrac{ak_m\nu_x E_2}{4} - \dfrac{ak_m G}{4} & \dfrac{aE_1}{2b} + \dfrac{abk_m^2 G}{6} & \\[2ex] \dfrac{ak_m\nu_x E_2}{4} + \dfrac{ak_m G}{4} & \dfrac{abk_m^2 E_2}{12} - \dfrac{aG}{2b} & -\dfrac{ak_m\nu_x E_2}{4} + \dfrac{ak_m G}{4} & \dfrac{abk_m^2 E_2}{6} + \dfrac{aG}{2b} \end{bmatrix}$$

$$E_1 = \frac{E_x}{1-\nu_x\nu_y}$$

$$E_2 = \frac{E_y}{1-\nu_x\nu_y}$$

Equation (3.11) has been used to work out the explicit form of the stiffness matrix for the lower order strip (Table 3.1). Note that the matrix coefficient expressions listed there are really only valid for the particular formulation in which $Y_m$ is used for $u$ and $Y'_m$ for $v$. For the other clamped–clamped case given by (3.5), some modifications in the expressions will be necessary.

The stiffness matrix listed in Table 3.1 is valid for all six types of end conditions and is widely used in vibration analysis. For the special case of two opposite ends simply supported, $[S]_{mn}$ becomes zero for $m \neq n$, and the very simple stiffness matrix $[S]_{mm}$ is presented in Table 3.2.

The consistent load matrix is obtained through the use of (1.35) and (3.7). Due to the linear displacement assumption in the transverse direction the matrix for uniform load case is particularly simple and

$$\{F\}_m = \frac{b}{2} \left\{ \begin{array}{c} q^x \displaystyle\int_0^a Y_m\, dy \\[2ex] q_y \dfrac{a}{\mu_m} \displaystyle\int_0^a Y'_m\, dy \\[2ex] q_x \displaystyle\int_0^a Y_m\, dy \\[2ex] q_y \dfrac{a}{\mu_m} \displaystyle\int_0^a Y'_m\, dy \end{array} \right\} \tag{3.12}$$

where $q_x$ and $q_y$ are the load components in the $x$ and $y$ directions respectively.

### 3.2.2. HIGHER ORDER STRIP WITH ONE INTERNAL NODAL LINE (HO3)

Referring to Fig. 3.1b it is seen that only displacement values exist at the three nodal lines, and therefore the situation concurs with that of case (d) in Section 1.3.2. A suitable set of displacement functions for the strip (with two opposite ends simply supported) is given by eqn. (3.13), in which $k_m = m\pi/a$ and $C_1$, $C_2$, and $C_3$ are the shape functions listed in (1.9d).

By applying (3.8) to (3.13), the strain matrix $[B]_m$ (3.14) is obtained.

$$\{f\} = \begin{Bmatrix} u \\ v \end{Bmatrix} = \sum_{m=1}^{r} \begin{bmatrix} C_1 \sin k_m y & 0 & C_2 \sin k_m y & 0 & C_3 \sin k_m y & 0 \\ 0 & C_1 \cos k_m y & 0 & C_2 \cos k_m y & 0 & C_3 \cos k_m y \end{bmatrix}_m \begin{Bmatrix} u_1 \\ v_1 \\ u_2 \\ v_2 \\ u_3 \\ v_3 \end{Bmatrix}_m \tag{3.13}$$

$$= \sum_{m=1}^{r} [N]_m \{\delta\}_m.$$

$$[B]_m = \begin{bmatrix} \dfrac{1}{b}(-3+4\bar{x})\sin k_m y & 0 & \dfrac{1}{b}(4-8\bar{x})\sin k_m y & 0 & \dfrac{1}{b}(-1+4\bar{x})\sin k_m y & 0 \\[2mm] 0 & -(1-3\bar{x}+2\bar{x}^2)k_m \sin k_m y & 0 & -(4\bar{x}-4\bar{x}^2)k_m \sin k_m y & 0 & -(-\bar{x}+2\bar{x}^2)k_m \sin k_m y \\[2mm] (1-3\bar{x}+2\bar{x}^2)k_m \cos k_m y & \dfrac{1}{b}(-3+4\bar{x})\cos k_m y & (4\bar{x}-4\bar{x}^2)k_m \cos k_m y & \dfrac{1}{b}(4-8\bar{x})\cos k_m y & (-\bar{x}+2\bar{x}^2)k_m \cos k_m y & \dfrac{1}{b}(-1+4\bar{x})\cos k_m y \end{bmatrix} \tag{3.14}$$

TABLE 3.3. STIFFNESS MATRIX FOR A HO3 STRIP FOR IN-PLANE STRESS ANALYSIS (LOO[5])

$$[S]_{mm} = \begin{bmatrix} K1 & & & & & \\ K2 & K7 & & \text{Symmetrical} & & \\ K3 & -K4 & K11 & & & \\ K4 & K8 & 0 & K12 & & \\ K5 & -K6 & K3 & -K4 & K1 & \\ K6 & K10 & K4 & K8 & -K2 & K7 \end{bmatrix}$$

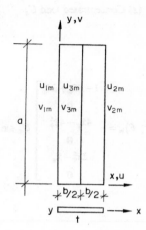

where,

$$K1 = \frac{7a}{6b} A + \frac{abk_m^2}{15} D$$

$$K2 = \frac{ak_m}{4} C - \frac{ak_m}{4} D$$

$$K3 = -\frac{4a}{3b} A + \frac{abk_m^2}{30} D$$

$$K4 = \frac{ak_m}{3} C + \frac{ak_m}{3} D$$

$$K5 = \frac{a}{6b} A - \frac{abk_m^2}{60} D$$

$$K6 = \frac{-ak_m}{12} C - \frac{ak_m}{12} D$$

$$K7 = \frac{abk_m^2}{15} B + \frac{7a}{6b} D$$

$$K8 = \frac{abk_m^2}{30} B - \frac{4a}{3b} D$$

$$K10 = -\frac{abk_m^2}{60} B + \frac{a}{6b} D$$

$$K11 - \frac{8a}{3b} A + \frac{4abk_m^2}{15} D$$

$$K12 = \frac{4abk_m^2}{15} B + \frac{8a}{3b} D$$

in which

$$A = \frac{E_x}{1 - v_x v_y} \qquad C = v_x A = v_y B$$

$$B = \frac{E_y}{1 - v_x v_y} \qquad D = G_{xy}$$

TABLE 3.4. FORCE MATRICES DUE TO CONCENTRATED IN-PLANE LOADS FOR A HO3 STRIP (LOO[5])

(a) Concentrated load $U_0$

$$\{F\}_m = \begin{bmatrix} 1 - 3\bar{x}_0 + 2\bar{x}_0^2 \\ 0 \\ 4\bar{x}_0 - 4\bar{x}_0^2 \\ 0 \\ 2\bar{x}_0^2 - \bar{x}_0 \\ 0 \end{bmatrix} U_0 \sin k_m y_0$$

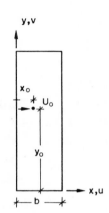

(b) Concentrated load $V_0$

$$\{F\}_m = \begin{bmatrix} 0 \\ 1 - 3\bar{x}_0 + 2\bar{x}_0^2 \\ 0 \\ 4\bar{x}_0 - 4\bar{x}_0^2 \\ 0 \\ 2\bar{x}_0^2 - \bar{x}_0 \end{bmatrix} V_0 \cos k_m y_0$$

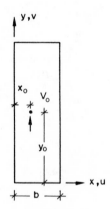

$$\bar{x}_0 = \frac{x_0}{b}$$

The stiffness matrix $[S]_{mm}$ is given in Table 3.3, while the consistent load matrices for several types of loading have also been worked out and are listed in Table 3.4. As discussed previously, from a computation point of view it is best to eliminate the parameters associated with the internal nodal line through static condensation, and the resulting reduced matrices have also been worked out by Loo.[5]

### 3.2.3. HIGHER ORDER STRIP WITH TWO NODAL LINES (HO2)

In this strip (Fig. 3.1c), a higher order displacement function is arrived at through the use of four (instead of two for the strips presented previously) parameters at each of the two nodal lines. The parameters involved are $u$, $\partial u/\partial x$, $v$, $\partial v/\partial x$, and therefore the relevant shape functions are supplied through (1.9b). The displacement function for the strip can now be written as in eqn. (3.15).

The formulation procedure given in Section 1.4.1 should now be followed to work out the stiffness and load matrices now that the displacement functions have been established.

### 3.3. CURVED STRIP (LO2)

In Section 2.3 it was shown how a curved bending strip can be evolved from a rectangular bending strip by transforming all the quantities associated with the Cartesian coordinate system into corresponding quantities associated with the polar coordinate system, and the same technique can be applied here in developing a curved plane stress strip. Such a curved strip has probably no direct application in actual two-dimensional engineering problems, but is mainly used in combination with a curved bending strip for analysing curved box beams of stiffened plates, thus still serving a very important purpose.

A lower order strip (Fig. 3.1d) was proposed by Cheung and Cheung[6, 7] in which, again, a linear variation of displacements across the section was assumed. The displacement functions are simple transformation of (3.7) and are written as in eqn. (3.16), in which $R$, $\alpha$, and $\Theta_m$ have already been defined in Section 2.3.

$$\{f\} = \begin{Bmatrix} u \\ v \end{Bmatrix} = \sum_{m=1}^{r} \begin{bmatrix} [C_1]\sin k_m y & 0 & [C_2]\sin k_m y & 0 \\ 0 & [C_1]\cos k_m y & 0 & [C_2]\cos k_m y \end{bmatrix} \underbrace{\begin{Bmatrix} u_1 \\ \left(\dfrac{\partial u}{\partial x}\right)_1 \\ v_1 \\ \left(\dfrac{\partial v}{\partial x}\right)_1 \\ u_2 \\ \left(\dfrac{\partial u}{\partial x}\right)_2 \\ v_2 \\ \left(\dfrac{\partial v}{\partial x}\right)_2 \end{Bmatrix}}_{m} = \sum_{m=1}^{r} [N]_m \{\delta\}_m . \tag{3.15}$$

$$\{f\} = \begin{Bmatrix} u \\ v \end{Bmatrix} = \sum_{m=1}^{r} \begin{bmatrix} \left(1-\dfrac{R}{2}\right)\Theta_m & 0 & \dfrac{R}{2}\,\Theta_m & 0 \\ 0 & \left(1-\dfrac{R}{2}\right)\dfrac{\alpha}{\mu_m}\,\Theta'_m & 0 & \dfrac{R}{2}\dfrac{\alpha}{\mu_m}\,\Theta'_m \end{bmatrix} \begin{Bmatrix} u_1 \\ v_1 \\ u_2 \\ v_2 \end{Bmatrix}_m \tag{3.16}$$

The strain–displacement relationship[8] is defined by

$$\{\varepsilon\} = \begin{Bmatrix} \varepsilon_r \\ \varepsilon_\theta \\ \gamma_{r\theta} \end{Bmatrix} = \begin{Bmatrix} \dfrac{\partial u}{\partial r} \\[2mm] \dfrac{1}{r}\dfrac{\partial v}{\partial \theta} + \dfrac{u}{r} \\[2mm] \dfrac{1}{r}\dfrac{\partial u}{\partial \theta} + \dfrac{\partial v}{\partial r} \end{Bmatrix} = \sum_{m=1}^{r} [B]_m \{\delta\}_m \qquad (3.17)$$

and the property matrix by

$$[D] = \begin{bmatrix} \dfrac{E_r}{1-v_\theta v_r} & \dfrac{v_\theta E_r}{1-v_\theta v_r} & 0 \\[3mm] \dfrac{v_\theta E_r}{1-v_\theta v_r} & \dfrac{E_\theta}{1-v_\theta v_r} & 0 \\[3mm] 0 & 0 & G_{r\theta} \end{bmatrix} \qquad (3.18)$$

where $E_r$, $E_\theta$, $v_r$, $v_\theta$, and $G_{r\theta}$ are the elastic constants for the polar coordinate system.

The stiffness matrix $[S]_{mn}$ for a constant thickness curved strip is computed through an equation similar to (3.11), i.e.

$$[S]_{mn} = t \int_0^\alpha \int_{r_1}^{r_2} [B]_m^T [D] [B]_n r\, d\theta\, dr. \qquad (3.19)$$

The expressions to be integrated in (3.19) are fairly complex and, therefore, to work out all the integrals in closed form is time-consuming. Furthermore, due to the presence of $1/r^n$ terms, such closed form integrals are numerically sensitive to very small or very large radii values, and a numerical integration scheme such as Gaussian quadrature should be used instead. In Table 3.5, $[S]_{mn}$ is shown in an explicit form, but with the integration step yet to be completed.

## 3.4. NUMERICAL EXAMPLES

In order to demonstrate the accuracy of the finite strip method in plane stress analysis, a number of deep beam and ordinary beam problems have been analysed using a LO2 rectangular strip.

TABLE 3.5. IN-PLANE STIFFNESS

$$[S]_{mn} = \int_{r_1}^{r_2}$$

| | |
|---|---|
| $+K_r\left(\dfrac{1}{4b'^2}\right)Y_mY_n$ <br> $-K_1\left(\dfrac{1}{rb'}\right)\left(1-\dfrac{R}{2}\right)Y_mY_n$ <br> $+K_\theta\left(\dfrac{1}{r}\right)^2\left(1-\dfrac{R}{2}\right)^2 Y_mY_n$ <br> $+K_{r\theta}\left(\dfrac{1}{r}\right)^2\left(1-\dfrac{R}{2}\right)^2 Y'_mY'_n$ | $-K_1\left(\dfrac{1}{2rb'}\right)C_n\left(1-\dfrac{R}{2}\right)Y_mY''_n$ <br> $+K_\theta\left(\dfrac{1}{r}\right)^2 C_n\left(1-\dfrac{R}{2}\right)^2 Y_mY''_n$ <br> $-K_{r\theta}\left(\dfrac{1}{2rb'}\right)C_n\left(1-\dfrac{R}{2}\right)Y'_mY'_n$ <br> $-K_{r\theta}\left(\dfrac{1}{r}\right)^2 C_n\left(1-\dfrac{R}{2}\right)Y'_mY'_n$ |
| $-K_1\left(\dfrac{1}{2rb'}\right)C_m\left(1-\dfrac{R}{2}\right)Y''_mY_n$ <br> $+K_\theta\left(\dfrac{1}{r}\right)^2 C_m\left(1-\dfrac{R}{2}\right)^2 Y''_mY_n$ <br> $-K_{r\theta}\left(\dfrac{1}{2rb'}\right)C_m\left(1-\dfrac{R}{2}\right)Y'_mY'_n$ <br> $-K_{r\theta}\left(\dfrac{1}{r}\right)^2 C_m\left(1-\dfrac{R}{2}\right)Y'_mY'_n$ | $+K_\theta\left(\dfrac{1}{r}\right)^2 C_mC_n\left(1-\dfrac{R}{2}\right)^2 Y''_mY''_n$ <br> $+K_{r\theta}\left(\dfrac{1}{2b'}\right)^2 C_mC_nY'_mY'_n$ <br> $+K_{r\theta}\left(\dfrac{1}{rb'}\right)C_mC_n\left(1-\dfrac{R}{2}\right)Y'_mY'_n$ <br> $+K_{r\theta}\left(\dfrac{1}{r}\right)^2 C_mC_n\left(1-\dfrac{R}{2}\right)^2 Y'_mY'_n$ |
| $-K_r\left(\dfrac{1}{2b'}\right)^2 Y_mY_n$ <br> $+K_1\left(\dfrac{1}{2rb'}\right)(1-R)Y_mY_n$ <br> $+K_\theta\left(\dfrac{1}{r}\right)^2\left(\dfrac{R}{2}\right)\left(1-\dfrac{R}{2}\right)Y_mY_n$ <br> $+K_{r\theta}\left(\dfrac{1}{r}\right)^2\left(\dfrac{R}{2}\right)\left(1-\dfrac{R}{2}\right)Y'_mY'_n$ | $+K_1\left(\dfrac{1}{2rb'}\right)C_n\left(1-\dfrac{R}{2}\right)Y_mY''_n$ <br> $+K_\theta\left(\dfrac{1}{r}\right)^2 C_n\left(\dfrac{R}{2}\right)\left(1-\dfrac{R}{2}\right)Y_mY''_n$ <br> $-K_{r\theta}\left(\dfrac{1}{2rb'}\right)C_n\left(\dfrac{R}{2}\right)Y'_mY'_n$ <br> $-K_{r\theta}\left(\dfrac{1}{r}\right)^2 C_n\left(\dfrac{R}{2}\right)\left(1-\dfrac{R}{2}\right)Y'_mY'_n$ |
| $-K_1\left(\dfrac{1}{2rb'}\right)C_m\left(\dfrac{R}{2}\right)Y''_mY_n$ <br> $+K_\theta\left(\dfrac{1}{r}\right)^2 C_m\left(\dfrac{R}{2}\right)\left(1-\dfrac{R}{2}\right)Y''_mY_n$ <br> $+K_{r\theta}\left(\dfrac{1}{2rb'}\right)C_m\left(1-\dfrac{R}{2}\right)Y'_mY'_n$ <br> $-K_{r\theta}\left(\dfrac{1}{r}\right)^2 C_m\left(\dfrac{R}{2}\right)\left(1-\dfrac{R}{2}\right)Y'_mY'_n$ | $+K_\theta\left(\dfrac{1}{r}\right)^2 C_mC_n\left(\dfrac{R}{2}\right)\left(1-\dfrac{R}{2}\right)Y''_mY_n$ <br> $-K_{r\theta}\left(\dfrac{1}{2b'}\right)^2 C_mC_nY'_mY'_n$ <br> $+K_{r\theta}\left(\dfrac{1}{2rb'}\right)C_mC_n(R-1)Y'_mY'_n$ <br> $+K_{r\theta}\left(\dfrac{1}{r}\right)^2 C_mC_n\left(\dfrac{R}{2}\right)\left(1-\dfrac{R}{2}\right)Y'_mY_n$ |

$$K_r = \frac{E_r}{1-\nu_r\nu_\theta}, \quad K_1 = \frac{\nu_r E_\theta}{1-\nu_r\nu_\theta}, \quad K_\theta = \frac{E_\theta}{1-\nu_r\nu_\theta},$$

$$Y_mY_n = \int_0^\alpha \Theta_m\Theta_n\,d\theta, \quad Y_mY''_n = \int_0^\alpha \Theta_m\Theta''_n\,d\theta,$$

$$
\left[
\begin{array}{c|c}
\begin{aligned}
&-K_r\left(\frac{1}{2b'}\right)^2 Y_m Y_n \\
&+K_1\left(\frac{1}{2rb'}\right)(1-R)Y_m Y_n \\
&+K_\theta\left(\frac{1}{r}\right)^2\left(\frac{R}{2}\right)\left(1-\frac{R}{2}\right)Y_m Y_n \\
&+K_{r\theta}\left(\frac{1}{r}\right)^2\left(\frac{R}{2}\right)\left(1-\frac{R}{2}\right)Y'_m Y'_n
\end{aligned}
&
\begin{aligned}
&-K_1\left(\frac{1}{2rb'}\right)C_n\left(\frac{R}{2}\right)Y_m Y''_n \\
&+K_\theta\left(\frac{1}{r}\right)^2 C_n\left(\frac{R}{2}\right)\left(1-\frac{R}{2}\right)Y_m Y''_n \\
&+K_{r\theta}\left(\frac{1}{2rb'}\right)C_n\left(1-\frac{R}{2}\right)Y'_m Y'_n \\
&-K_{r\theta}\left(\frac{1}{r}\right)^2 C_n\left(\frac{R}{2}\right)\left(1-\frac{R}{2}\right)Y'_m Y'_n
\end{aligned}
\\ \hline
\begin{aligned}
&+K_1\left(\frac{1}{2rb'}\right)C_m\left(1-\frac{R}{2}\right)Y''_m Y_n \\
&+K_\theta\left(\frac{1}{r}\right)^2 C_m\left(\frac{R}{2}\right)\left(1-\frac{R}{2}\right)Y''_m Y_n \\
&-K_{r\theta}\left(\frac{1}{2rb'}\right)C_m\left(\frac{R}{2}\right)Y'_m Y'_n \\
&-K_{r\theta}\left(\frac{1}{r}\right)^2 C_m\left(\frac{R}{2}\right)\left(1-\frac{R}{2}\right)Y'_m Y'_n
\end{aligned}
&
\begin{aligned}
&+K_\theta\left(\frac{1}{r}\right)^2 C_m C_n\left(\frac{R}{2}\right)\left(1-\frac{R}{2}\right)Y''_m Y''_n \\
&-K_{r\theta}\left(\frac{1}{2b'}\right)^2 C_m C_n Y'_m Y'_n \\
&+K_{r\theta}\left(\frac{1}{2rb'}\right)C_m C_n(R-1)Y'_m Y'_n \\
&+K_{r\theta}\left(\frac{1}{r}\right)^2 C_m C_n\left(\frac{R}{2}\right)\left(1-\frac{R}{2}\right)Y'_m Y'_n
\end{aligned}
\\ \hline
\begin{aligned}
&+K_r\left(\frac{1}{2b'}\right)^2 Y_m Y_n \\
&+K_1\left(\frac{1}{rb'}\right)\left(\frac{R}{2}\right)Y_m Y_n \\
&+K_\theta\left(\frac{1}{r}\right)^2\left(\frac{R}{2}\right)Y_m Y_n \\
&+K_{r\theta}\left(\frac{1}{r}\right)^2\left(\frac{R}{2}\right)^2 Y'_m Y'_n
\end{aligned}
&
\begin{aligned}
&+K_1\left(\frac{1}{2rb'}\right)C_n\left(\frac{R}{2}\right)Y_m Y''_n \\
&+K_\theta\left(\frac{1}{r}\right)^2 C_n\left(\frac{R}{2}\right)^2 Y_m Y''_n \\
&+K_{r\theta}\left(\frac{1}{2rb'}\right)C_n\left(\frac{R}{2}\right)Y'_m Y'_n \\
&-K_{r\theta}\left(\frac{1}{r}\right)^2 C_n\left(\frac{R}{2}\right)^2 Y'_m Y'_n
\end{aligned}
\\ \hline
\begin{aligned}
&+K_1\left(\frac{1}{2rb'}\right)C_m\left(\frac{R}{2}\right)Y''_m Y_n \\
&+K_\theta\left(\frac{1}{r}\right)^2 C_m\left(\frac{R}{2}\right)^2 Y''_m Y_n \\
&+K_{r\theta}\left(\frac{1}{2rb'}\right)C_m\left(\frac{R}{2}\right)Y'_m Y'_n \\
&-K_{r\theta}\left(\frac{1}{r}\right)^2 C_m\left(\frac{R}{2}\right)^2 Y'_m Y'_n
\end{aligned}
&
\begin{aligned}
&+K_\theta\left(\frac{1}{r}\right)^2 C_m C_n\left(\frac{R}{2}\right)^2 Y''_m Y''_n \\
&+K_{r\theta}\left(\frac{1}{2b'}\right)^2 C_m C_n Y'_m Y'_n \\
&-K_{r\theta}\left(\frac{1}{rb'}\right)C_m C_n\left(\frac{R}{2}\right)Y'_m Y'_n \\
&+K_{r\theta}\left(\frac{1}{r}\right)^2 C_m C_n\left(\frac{R}{2}\right)^2 Y'_m Y'_n
\end{aligned}
\end{array}
\right]\, r\,dr
$$

$$
K_{r\theta} = G_{r\theta}, \quad R = \frac{r-r_1}{b'}, \quad b' = \frac{r_2 - r_1}{2},
$$

$$
Y''_m Y''_n = \int_0^\alpha \Theta''_m \Theta''_n \, d\theta, \quad C_m = \frac{\alpha}{\mu_m}, \quad C_n = \frac{\alpha}{\mu_n}
$$

The first example deals with the analysis of an isotropic, simply supported square deep beam under uniform line load acting at the top, and the
results are presented in Fig. 3.2. It is found that both the longitudinal
stress $\sigma_y$ and transverse stress $\sigma_x$ agree very well with the values given
by a finite element analysis ($16 \times 16$ mesh of constant strain triangular

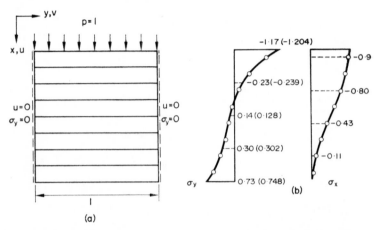

FIG. 3.2. (a) Simply supported deep beam and its strip idealization. (b)
Stress distributions at central section of the deep beam. $16 \times 16$ finite
element solution; o o o finite strip solution; ( ) Kalmanok.[9]

elements). The stress $\sigma_y$ is also compared with the values given in a  book
Kalmanok,[9] and again very good agreement is demonstrated.

The transverse stress $\sigma_x$ has been plotted at the middle of each strip
in order to avoid a stepwise type of representation. For the longitudinal
stress $\sigma_y$ the plotting location is immaterial because of the very small
jump at a nodal line.

The second example demonstrates the use of other types of series other
than the particular type used for simply supported conditions. The same
deep beam with clamped ends is now analysed using the displacement
function given in (3.5). The results are presented in Fig. 3.3, and once
more the comparisons with a $16 \times 16$ finite element analysis results are excellent.

Finally, a simple beam[5] subject to two asymmetric longitudinal concentrated loads at the two end supports is analysed, using ten equal strips.

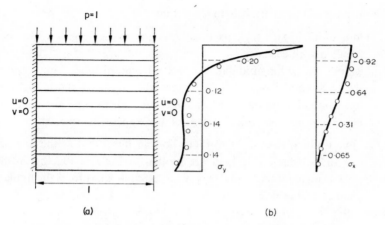

FIG. 3.3. (a) A clamped deep beam and its strip idealization. (b) Stress distributions at central section of the deep beam — 16×16 finite element solution. o o o finite strip solution.

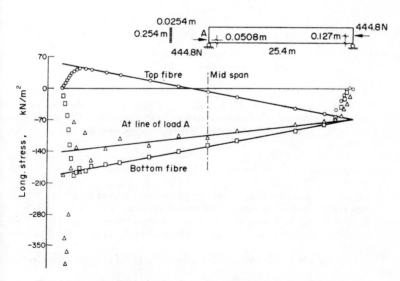

FIG. 3.4. Longitudinal distribution of longitudinal stresses for a simple beam under asymmetrical end anchorage forces (Loo[5]). Simple beam theory. O, □, △ 10 LO2 strips and 49 harmonics.

The purpose of this analysis is to show that:

(i) longitudinal loads such as those due to prestressing can also be handled with ease;

(ii) as long as the longitudinal forces are balanced when considering the gross equilibrium of the structure (due to the condition that no longitudinal stress can exist at the two ends), they do not have to act along the same line of action as was sometimes believed.

The longitudinal stresses at top and bottom fibres and at the line of action of the left-hand-side load are plotted in Fig. 3.4 together with the simple beam analysis results. It may be seen that apart from the immediate vicinity of the point load, the strip analysis and the simple beam theory give very good correlation. The beam analysis of course does not give any idea of the actual stress distribution at the two ends where stress concentrations are present.

The last example is of some practical significance because it demonstrates beyond any doubt that prestressing loading, which is very common in box girder bridge analysis, can be handled without difficulty.

## REFERENCES

1. Y. C. FUNG, *Foundation of Solid Mechanics*, Prentice-Hall, 1965.
2. Y. K. CHEUNG, Analysis of box girder bridges by finite strip method, *Second International Symposium on Concrete Bridge Design, Chicago, March, 1969*. Also *Concrete Bridge Design*, ACI Publications SP-26, pp. 357–78, 1971.
3. Y. C. LOO and A. R. CUSENS, The auxiliary nodal line technique for the analysis of box girder bridge decks by the finite strip method, *Proceedings of Specialty Conference on the Finite Element Method in Civil Engineering, McGill University, Montreal, June 1972*.
4. K. YOSHIDA and N. OKA, *A Note on the In-plane Displacement Functions of Strip Elements*, Research Report, Department of Naval Architecture, Tokyo University, 1972.
5. Y. C. LOO, Developments and applications of the finite strip method in the analysis of right bridge decks, PhD thesis, University of Dundee, 1972.
6. M. S. CHEUNG and Y. K. CHEUNG, Analysis of curved box girder bridges by finite strip method, *Publications, International Association for Bridges and Structural Engineering* 31-I, 1971.
7. M. S. CHEUNG, Finite strip analysis of structures, PhD thesis, University of Calgary, 1971.
8. S. P. TIMOSHENKO and S. WOINOWSKY-KRIEGER, *Theory of Elasticity*, 2nd edn., McGraw-Hill, 1951.
9. A. S. KALMANOK, *Calculation of Plates*, National Press, Moscow, 1959 (in Russian).

# CHAPTER 4

# *Analysis of folded plate structures with special reference to box girder bridges*

## 4.1. INTRODUCTION

The analysis of folded plate structures has been the centre of attention for many researchers, and much information on this topic can be found in an ASCE report.[1] The elasticity method, which was developed by Goldberg and Leve[2] and subsequently programmed and applied by De Fries-Skene and Scordelis[3] as a stiffness approach, became quite popular because of its accuracy. The method however suffers from being fairly complex, and is difficult to apply to orthotropic folded plate structures and to dynamic analysis. The elasticity method has also been used in the study of simply supported box girder bridges by Scordelis[4] and by Chu and Dudnik.[5]

Recently, the elasticity method has been extended to cover the analysis of curved box girder bridges,[6-8] but the disadvantages outlined above also exist for this case.

The finite strip analysis of prismatic folded plate structures and box girder bridges was first introduced by Cheung[9, 10] who combined together a third-order bending strip and a linear plane stress strip to form a shell strip with both bending and membrane actions. The lower order shell strip was then applied to the study of orthotropic folded plates[11] and eccentrically folded plates.[12] A higher order flat shell strip with one internal nodal line was later developed by Loo and Cusens[13] and applied to the analysis of right box girder bridge decks. Finally, solutions for curved folded plates and box girder bridges were presented by Meyer and Scordelis,[14] and independently by Cheung and Cheung,[15] in which a conical frustrum shell strip was used.

The basic assumptions used in the present analysis and outlined in the following paragraphs are for prismatic folded plates. For convenience, remarks relevant to curved structures are given in parenthesis:

(i) The webs and flanges are made up of isotropic or (cylindrical) orthotropic materials.

(ii) The folded plate structure is bounded by two straight lines (two concentric circular arcs) and by two (radial) planes at the supports.

(iii) The plates are supported at each end by a diaphragm which is infinitely stiff in its own plane but infinitely flexible out of plane. An idealized simply supported condition is thus achieved both for in-plane and out of plane behaviours.

(iv) The simply supported conditions allow the different terms of the series to decouple and each term can therefore be computed separately. All future discussions will henceforth be limited to a typical $m$th term.

The above assumptions are of course also valid for simply supported box girder bridges.

## 4.2. FORMULATION OF STIFFNESS MATRIX

### 4.2.1. RECTANGULAR STRIPS (FIG. 4.1)

For a rectangular flat shell strip it is assumed that no interaction takes place between the bending and membrane actions that exist in a folded plate structure. As a result, a shell strip can be formed through a simple combination of a bending strip (Section 2.2.1 or 2.2.3) and a plane stress strip (Section 3.2.1 or 3.2.2) in the following manner.

Let both bending and in-plane nodal displacements act simultaneously, then at each of the two nodal lines there will be four displacement components and four corresponding force components. The displacement components are related to the force components through a stiffness matrix. In this context, for a LO2 strip with only nodal lines 1 and 2, it is possible to write

$$\begin{bmatrix} S_{11} & S_{12} \\ S_{21} & S_{22} \end{bmatrix}_{mm} \begin{Bmatrix} \{\delta_1\} \\ \{\delta_2\} \end{Bmatrix}_m = \begin{Bmatrix} \{F_1\} \\ \{F_2\} \end{Bmatrix}_m \qquad (4.1a)$$

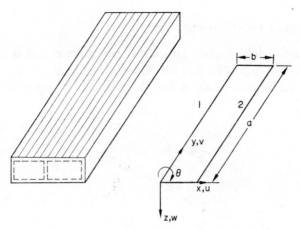

FIG. 4.1. A typical bridge and a rectangular shell strip.

or

$$[S]_{mm}\{\delta\}_m = \{F\}_m \tag{4.1b}$$

in which

$$\{\delta\}_m = [u_1 v_1 w_1 \theta_1 u_2 v_2 w_2 \theta_2]_m^T, \tag{4.1c}$$

$$\{F\}_m = [U_1 V_1 W_1 M_1 U_2 V_2 W_2 M_2]_m^T. \tag{4.1d}$$

A typical submatrix $[S_{ij}]_{mm}$ of the matrix $[S]_{mm}$ is made up from ape propriate submatrices of the bending stiffness matrix $[S_{ij}^b]_{n:m}$ and plan-stress stiffness matrix $[S_{ij}^p]_{mm}$. Thus

$$\underset{(4\times4)}{[S_{ij}]_{mm}} = \begin{bmatrix} [S_{ij}^p]_{mm} & [0] \\ (2\times2) & \\ [0] & [S_{ij}^b]_{mm} \\ & (2\times2) \end{bmatrix}. \tag{4.2}$$

The above formulation is generally valid for rectangular flat shell strips and applies to both lower order and higher order strips.

For a LO2 strip, $[S^b]_{mm}$ and $[S^p]_{mm}$ are given in Tables 2.2 and 3.2 respectively, while for a HO3 strip with one internal nodal line, the corresponding matrices can be found in Tables 2.5 and 3.3.

## 4.2.2. CONICAL FRUSTRUM SHELL STRIP

For curved folded plate structures the common unit is a conical frustrum shell strip (Fig. 4.2) which, however, changes to a cylindrical strip in the vertical position (for $\phi = 0°$), and to a flat sector strip

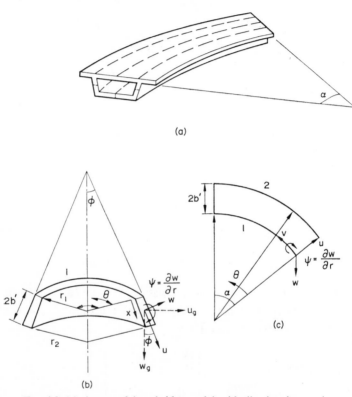

(a)

(b)

(c)

FIG. 4.2. (a) A curved box bridge and its idealization into strips. (b) A conical web strip. (c) A flange strip.

when horizontal (for $\phi = 90°$). Since the strip is no longer flat in this case, the bending and membrane forces are now coupled and thus a simple formulation similar to the one presented in the previous section is no longer available.

Referring to Fig. 4.2, the displacement functions for the $m$th term of the series are:

TABLE 4.1. STRAIN MATRIX OF A CONICAL WEB STRIP

$[B]_m =$

| | | | | | | | |
|---|---|---|---|---|---|---|---|
| $-\dfrac{1}{b}S$ | $0$ | $0$ | $0$ | $\dfrac{1}{b}S$ | $0$ | $0$ | $0$ |
| $\dfrac{1}{r}\left(1-\dfrac{x}{b}\right)SS\phi$ | $-\dfrac{1}{r}\left(1-\dfrac{x}{b}\right)k_m S$ | $\dfrac{1}{r}\left(1-\dfrac{3x^2}{b^2}+\dfrac{2x^3}{b^3}\right)SC\phi$ | $\dfrac{1}{r}\left(z-\dfrac{2x^2}{b}+\dfrac{x^3}{b^2}\right)SC\phi$ | $\dfrac{1}{r}\left(\dfrac{x}{b}\right)SS\phi$ | $-\left(\dfrac{x}{rb}\right)k_m S$ | $\dfrac{1}{r}\left(\dfrac{3x^2}{b^2}-\dfrac{2x^3}{b^3}\right)SC\phi$ | $\dfrac{1}{r}\left(\dfrac{x^3}{b^2}-\dfrac{x^2}{b}\right)SC\phi$ |
| $\dfrac{1}{r}\left(1-\dfrac{x}{b}\right)k_m C$ | $-\dfrac{1}{b}C$ $\dfrac{1}{r}\left(1-\dfrac{x}{b}\right)CS\phi$ | $0$ | $0$ | $\dfrac{1}{r}\left(\dfrac{x}{b}\right)k_m C$ | $\dfrac{1}{b}C$ $-\dfrac{1}{r}\left(\dfrac{x}{b}\right)CS\phi$ | $0$ | $0$ |
| $0$ | $0$ | $\left(\dfrac{6}{b^2}-\dfrac{12x}{b^3}\right)S$ | $\left(\dfrac{4}{b}-\dfrac{6x}{b^2}\right)S$ | $0$ | $0$ | $\left(\dfrac{-6}{b^2}+\dfrac{12x}{b^3}\right)S$ | $\left(\dfrac{-6x}{b^2}+\dfrac{2}{b}\right)S$ |
| $0$ | $-\dfrac{1}{r^2}\left(1-\dfrac{x}{b}\right)k_m SC\phi$ | $\dfrac{1}{r^2}\left(1-\dfrac{3x^2}{b^2}+\dfrac{2x^3}{b^3}\right)k_m^2 S$ $-\dfrac{1}{r}\left(-\dfrac{6x}{b^2}+\dfrac{6x^2}{b^3}\right)SS\phi$ | $\dfrac{1}{r^2}\left(x-\dfrac{2x^2}{b}+\dfrac{x^3}{b^2}\right)k_m^2 S$ $-\dfrac{1}{r}\left(1-\dfrac{4x}{b}+\dfrac{3x^2}{b^2}\right)SS\phi$ | $0$ | $-\dfrac{1}{r^2}\left(\dfrac{x}{b}\right)k_m SC\phi$ | $\dfrac{1}{r^2}\left(\dfrac{3x^2}{b^2}-\dfrac{2x^3}{b^3}\right)k_m^2 S$ $-\dfrac{1}{r}\left(\dfrac{6x}{b^2}-\dfrac{6x^2}{b^3}\right)SS\phi$ | $\dfrac{1}{r}\left(\dfrac{x^3}{b^2}-\dfrac{x^2}{b}\right)k_m^2 S$ $-\dfrac{1}{r^2}\left(\dfrac{3x^2}{b^2}-\dfrac{2x}{b}\right)SS\phi$ |
| $0$ | $-\dfrac{2}{r}\left(\dfrac{1}{b}\right)CC\phi$ $-\dfrac{2}{r^2}\left(1-\dfrac{x}{b}\right)CS\phi C\phi$ | $\dfrac{2}{r}\left(\dfrac{6x}{b^2}-\dfrac{6x^2}{b^3}\right)k_m C$ $+\dfrac{2}{r^2}\left(1-\dfrac{3x^2}{b^2}+\dfrac{2x^3}{b^3}\right)k_m CS\phi$ | $\dfrac{2}{r}\left(-1+\dfrac{4x}{b}-\dfrac{3x^2}{b^2}\right)k_m C$ $+\dfrac{2}{r^2}\left(x-\dfrac{2x^2}{b}+\dfrac{x^3}{b}\right)k_m CS\phi$ | $0$ | $\dfrac{2}{r}\left(\dfrac{1}{b}\right)CC\phi$ $-\dfrac{2}{r^2}\left(\dfrac{x}{b}\right)CS\phi C\phi$ | $\dfrac{2}{r}\left(\dfrac{6x^2}{b^3}-\dfrac{6x}{b^2}\right)k_m C$ $+\dfrac{2}{r^2}\left(\dfrac{3x^2}{b^2}-\dfrac{2x^3}{b^3}\right)k_m CS\phi$ | $\dfrac{2}{r}\left(\dfrac{2x}{b}-\dfrac{3x^2}{b^2}\right)k_m C$ $+\dfrac{2}{r^2}\left(\dfrac{x^3}{b^2}-\dfrac{x^2}{b}\right)k_m CS\phi$ |

$$\left(k_m=\frac{m\pi}{\alpha}, \quad S=\sin k_m\theta, \quad C=\cos k_m\theta\right), \quad S\phi=\sin\phi, \quad C\phi=\cos\phi, \quad b=2b'$$

$$
\left.\begin{aligned}
u_m &= [(1-\bar{x})\,u_{1m}+(\bar{x})\,u_{2m}]\,\sin\frac{m\pi\theta}{\alpha}, \\
v_m &= [(1-\bar{x})\,v_{1m}+(\bar{x})\,v_{2m}]\,\cos\frac{m\pi\theta}{\alpha}, \\
w_m &= [(1-3\bar{x}^2+2\bar{x}^3)\,w_{1m}+x(1-2\bar{x}+\bar{x}^2)\,\psi_{1m} \\
&\quad +(3\bar{x}^2-2\bar{x}^3)\,w_{2m}+x(\bar{x}^2-\bar{x})\,\psi_{2m}]\,\sin\frac{m\pi\theta}{\alpha},
\end{aligned}\right\} \tag{4.3}
$$

in which $\bar{x} = x/2b'$.

The strain displacement relationship for a conical shell is somewhat more complicated than what we have seen up to now and is given by

$$
\left\{\begin{array}{c}
\varepsilon_x \\
\varepsilon_\theta \\
\gamma_{x\theta} \\
\chi_x \\
\chi_\theta \\
\chi_{x\theta}
\end{array}\right\}_m
=
\left\{\begin{array}{c}
\dfrac{\partial u}{\partial x} \\[2ex]
\dfrac{1}{r}\dfrac{\partial v}{\partial\theta}+\dfrac{w\cos\phi+u\sin\phi}{r} \\[2ex]
\dfrac{1}{r}\dfrac{\partial u}{\partial\theta}+\dfrac{\partial v}{\partial x}-\dfrac{v\sin\phi}{r} \\[2ex]
-\dfrac{\partial^2 w}{\partial x^2} \\[2ex]
-\dfrac{1}{r^2}\dfrac{\partial^2 w}{\partial\theta^2}+\dfrac{\cos\phi}{r^2}\dfrac{\partial v}{\partial\theta}-\dfrac{\sin\phi}{r}\dfrac{\partial w}{\partial x} \\[2ex]
2\left(-\dfrac{1}{r}\dfrac{\partial^2 w}{\partial x\,\partial\theta}+\dfrac{\sin\phi}{r^2}\dfrac{\partial w}{\partial\theta}+\dfrac{\cos\phi}{r}\dfrac{\partial v}{\partial x}-\dfrac{\sin\phi\cos\phi}{r^2}v\right)
\end{array}\right\}_m
$$

$$
= [B]_m\{\delta\}_m \tag{4.4}
$$

in which $\{\delta\}_m$ has been defined in (4.1b).

The elasticity matrix for a general orthotropic material reads as follows:

$$
[D] = \left\{\begin{array}{cccccc}
K_x & K_2 & 0 & 0 & 0 & 0 \\
K_2 & K_\theta & 0 & 0 & 0 & 0 \\
0 & 0 & K_{x\theta} & 0 & 0 & 0 \\
0 & 0 & 0 & D_x & D_2 & 0 \\
0 & 0 & 0 & D_2 & D_\theta & 0 \\
0 & 0 & 0 & 0 & 0 & D_{x\theta}
\end{array}\right\} \tag{4.5}
$$

TABLE 4.2. STIFFNESS MATRIX OF A SHELL STRIP

$[S_{11}]_{mn} =$

| | | | |
|---|---|---|---|
| $K_x B_{11}^2 + 2K_2 B_{21}B_{11}$<br>$+ K_\theta B_{21}^2 + K_{x\theta}B_{31}^2$<br>$+ D_x B_{41}^2 + 2D_2 B_{51}B_{41}$<br>$+ D_\theta B_{51}^2 + D_{x\theta}B_{61}^2$ | $K_x B_{12}B_{11} + K_2 B_{22}B_{11}$<br>$+ K_2 B_{12}B_{21} + K_\theta B_{22}B_{21}$<br>$+ K_{x\theta}B_{32}B_{31} + D_x B_{42}B_{41}$<br>$+ D_2 B_{52}B_{41} + D_2 B_{42}B_{51}$<br>$+ D_\theta B_{52}B_{51} + D_{x\theta}B_{62}B_{61}$ | $K_x B_{13}B_{11} + K_2 B_{23}B_{11}$<br>$+ K_2 B_{13}B_{21} + K_\theta B_{23}B_{21}$<br>$+ K_{x\theta}B_{33}B_{31} + D_x B_{43}B_{41}$<br>$+ D_2 B_{53}B_{41} + D_2 B_{43}B_{51}$<br>$+ D_\theta B_{53}B_{51} + D_{x\theta}B_{63}B_{61}$ | $K_x B_{14}B_{11} + K_2 B_{24}B_{11}$<br>$+ K_2 B_{14}B_{21} + K_\theta B_{24}B_{21}$<br>$+ K_{x\theta}B_{34}B_{31} + D_x B_{44}B_{41}$<br>$+ D_2 B_{54}B_{41} + D_2 B_{44}B_{51}$<br>$+ D_\theta B_{54}B_{51} + D_{x\theta}B_{64}B_{61}$ |
| $K_x B_{11}B_{12} + K_2 B_{21}B_{12}$<br>$+ K_2 B_{11}B_{22} + K_\theta B_{21}B_{22}$<br>$+ K_{x\theta}B_{31}B_{32} + D_x B_{41}B_{42}$<br>$+ D_2 B_{51}B_{42} + D_2 B_{41}B_{52}$<br>$+ D_\theta B_{51}B_{52} + D_{x\theta}B_{61}B_{62}$ | $K_x B_{12}^2 + 2K_2 B_{22}B_{12}$<br>$+ K_\theta B_{22}^2 + K_{x\theta}B_{32}^2$<br>$+ D_x B_{42}^2 + 2D_2 B_{52}B_{42}$<br>$+ D_\theta B_{52}^2 + D_{x\theta}B_{62}^2$ | $K_x B_{13}B_{12} + K_2 B_{23}B_{12}$<br>$+ K_2 B_{13}B_{22} + K_\theta B_{23}B_{22}$<br>$+ K_{x\theta}B_{33}B_{32} + D_x B_{43}B_{42}$<br>$+ D_2 B_{53}B_{42} + D_2 B_{43}B_{52}$<br>$+ D_\theta B_{53}B_{52} + D_{x\theta}B_{63}B_{62}$ | $K_x B_{14}B_{12} + K_2 B_{24}B_{12}$<br>$+ K_2 B_{14}B_{22} + K_\theta B_{24}B_{22}$<br>$+ K_{x\theta}B_{34}B_{32} + D_x B_{44}B_{42}$<br>$+ D_2 B_{54}B_{42} + D_2 B_{44}B_{52}$<br>$+ D_\theta B_{54}B_{52} + D_{x\theta}B_{64}B_{62}$ |
| $K_x B_{11}B_{13} + K_2 B_{21}B_{13}$<br>$+ K_2 B_{11}B_{23} + K_\theta B_{21}B_{23}$<br>$+ K_{x\theta}B_{31}B_{33} + D_x B_{41}B_{43}$<br>$+ D_2 B_{51}B_{43} + D_2 B_{41}B_{53}$<br>$+ D_\theta B_{51}B_{53} + D_{x\theta}B_{61}B_{63}$ | $K_x B_{12}B_{13} + K_2 B_{22}B_{13}$<br>$+ K_2 B_{12}B_{23} + K_\theta B_{22}B_{23}$<br>$+ K_{x\theta}B_{32}B_{33} + D_x B_{42}B_{43}$<br>$+ D_2 B_{52}B_{43} + D_2 B_{42}B_{53}$<br>$+ D_\theta B_{52}B_{53} + D_{x\theta}B_{62}B_{63}$ | $K_x B_{13}^2 + 2K_2 B_{23}B_{13}$<br>$+ K_\theta B_{23}^2 + K_{x\theta}B_{33}^2$<br>$+ D_x B_{43}^2 + 2D_2 B_{53}B_{43}$<br>$+ D_\theta B_{53}^2 + D_{x\theta}B_{63}^2$ | $K_x B_{14}B_{13} + K_2 B_{24}B_{13}$<br>$+ K_2 B_{14}B_{23} + K_\theta B_{24}B_{23}$<br>$+ K_{x\theta}B_{34}B_{33} + D_x B_{44}B_{43}$<br>$+ D_2 B_{54}B_{43} + D_2 B_{44}B_{53}$<br>$+ D_\theta B_{54}B_{53} + D_{x\theta}B_{64}B_{63}$ |
| $K_x B_{11}B_{14} + K_2 B_{21}B_{14}$<br>$+ K_2 B_{11}B_{24} + K_\theta B_{21}B_{24}$<br>$+ K_{x\theta}B_{31}B_{34} + D_x B_{41}B_{44}$<br>$+ D_2 B_{51}B_{44} + D_2 B_{41}B_{54}$<br>$+ D_\theta B_{51}B_{54} + D_{x\theta}B_{61}B_{64}$ | $K_x B_{12}B_{14} + K_2 B_{22}B_{14}$<br>$+ K_2 B_{12}B_{24} + K_\theta B_{22}B_{24}$<br>$+ K_{x\theta}B_{32}B_{34} + D_x B_{42}B_{44}$<br>$+ D_2 B_{52}B_{44} + D_2 B_{42}B_{54}$<br>$+ D_\theta B_{52}B_{54} + D_{x\theta}B_{62}B_{64}$ | $K_x B_{13}B_{14} + K_2 B_{23}B_{14}$<br>$+ K_2 B_{13}B_{24} + K_\theta B_{23}B_{24}$<br>$+ K_{x\theta}B_{33}B_{34} + D_x B_{43}B_{44}$<br>$+ D_2 B_{53}B_{44} + D_2 B_{43}B_{54}$<br>$+ D_\theta B_{53}B_{54} + D_{x\theta}B_{63}B_{64}$ | $K_x B_{14}^2 + 2K_2 B_{24}B_{14}$<br>$+ K_\theta B_{24}^2 + K_{x\theta}B_{34}^2$<br>$+ D_x B_{44}^2 + 2D_2 B_{54}B_{44}$<br>$+ D_\theta B_{54}^2 + D_{x\theta}B_{64}^2$ |

$(B_{ij}B_{kl}$ refers to product of the $m$th term and $n$th term coefficients in the strain matrix).

$[\bar{S}_{21}]_{mn}^T =$
$[\bar{S}_{12}]_{mn} =$

| | | | |
|---|---|---|---|
| $K_x B_{15} B_{11} + K_2 B_{25} B_{11}$<br>$+ K_2 B_{15} B_{21} + K_\theta B_{25} B_{21}$<br>$+ K_{x\theta} B_{35} B_{31} + D_x B_{45} B_{41}$<br>$+ D_2 B_{55} B_{41} + D_2 B_{45} B_{51}$<br>$+ D_\theta B_{55} B_{51} + D_{x\theta} B_{65} B_{61}$ | $D_x B_{16} B_{11} + K_2 B_{26} B_{11}$<br>$+ K_2 B_{16} B_{21} + K_\theta B_{26} B_{21}$<br>$+ K_{x\theta} B_{36} B_{31} + D_x B_{46} B_{41}$<br>$+ D_2 B_{56} B_{41} + D_2 B_{46} B_{51}$<br>$+ D_\theta B_{56} B_{51} + D_{x\theta} B_{66} B_{61}$ | $K_x B_{17} B_{11} + K_2 B_{27} B_{11}$<br>$+ K_2 B_{17} B_{21} + K_\theta B_{27} B_{21}$<br>$+ K_{x\theta} B_{37} B_{31} + D_x B_{47} B_{41}$<br>$+ D_2 B_{57} B_{41} + D_2 B_{47} B_{51}$<br>$+ D_\theta B_{57} B_{51} + D_{x\theta} B_{67} B_{61}$ | $K_x B_{18} B_{11} + K_2 B_{28} B_{11}$<br>$+ K_2 B_{18} B_{21} + K_\theta B_{28} B_{21}$<br>$+ K_{x\theta} B_{38} B_{31} + D_x B_{48} B_{41}$<br>$+ D_2 B_{58} B_{41} + D_2 B_{48} B_{51}$<br>$+ D_\theta B_{58} B_{51} + D_{x\theta} B_{68} B_{61}$ |
| $K_x B_{15} B_{12} + K_2 B_{25} B_{12}$<br>$+ K_2 K_{15} B_{22} + K_\theta B_{25} B_{22}$<br>$+ K_{x\theta} B_{35} B_{32} + D_x B_{45} B_{42}$<br>$+ D_2 B_{55} B_{42} + D_2 B_{45} B_{52}$<br>$+ D_\theta B_{55} B_{52} + D_{x\theta} B_{65} B_{62}$ | $K_x B_{16} B_{12} + K_2 B_{26} B_{12}$<br>$+ K_2 B_{16} B_{22} + K_\theta B_{26} B_{22}$<br>$+ K_{x\theta} B_{36} B_{32} + D_x B_{46} B_{42}$<br>$+ D_2 B_{56} B_{42} + D_2 B_{46} B_{52}$<br>$+ D_\theta B_{56} B_{52} + D_{x\theta} B_{66} B_{62}$ | $K_x B_{17} B_{12} + K_2 B_{27} B_{12}$<br>$+ K_2 B_{17} B_{22} + K_\theta B_{27} B_{22}$<br>$+ K_{x\theta} B_{37} B_{32} + D_x B_{47} B_{42}$<br>$+ D_2 B_{57} B_{42} + D_2 B_{47} B_{52}$<br>$+ D_\theta B_{57} B_{52} + D_{x\theta} B_{67} B_{62}$ | $K_x B_{18} B_{12} + K_2 B_{28} B_{12}$<br>$+ K_2 B_{18} B_{22} + K_\theta B_{28} B_{22}$<br>$+ K_{x\theta} B_{38} B_{32} + D_x B_{48} B_{42}$<br>$+ D_2 B_{58} B_{42} + D_2 B_{48} B_{52}$<br>$+ D_\theta B_{58} B_{52} + D_{x\theta} B_{68} B_{62}$ |
| $K_x B_{15} B_{13} + K_2 B_{25} B_{13}$<br>$+ K_2 B_{15} B_{23} + K_\theta B_{25} B_{23}$<br>$+ K_{x\theta} B_{35} B_{33} + D_x B_{45} B_{43}$<br>$+ D_2 B_{55} B_{43} + D_2 B_{45} B_{53}$<br>$+ D_\theta B_{55} B_{53} + D_{x\theta} B_{65} B_{63}$ | $K_x B_{16} B_{13} + K_2 B_{26} B_{13}$<br>$+ K_2 B_{16} B_{23} + K_\theta B_{26} B_{23}$<br>$+ K_{x\theta} B_{36} B_{33} + D_x B_{46} B_{43}$<br>$+ D_2 B_{56} B_{43} + D_2 B_{46} B_{53}$<br>$+ D_\theta B_{56} B_{53} + D_{x\theta} B_{66} B_{63}$ | $K_x B_{17} B_{13} + K_2 B_{27} B_{13}$<br>$+ K_2 B_{17} B_{23} + K_\theta B_{27} B_{23}$<br>$+ K_{x\theta} B_{37} B_{33} + D_x B_{47} B_{43}$<br>$+ D_2 B_{57} B_{43} + D_2 B_{47} B_{53}$<br>$+ D_\theta B_{57} B_{53} + D_{x\theta} B_{67} B_{63}$ | $K_x B_{18} B_{13} + K_2 B_{28} B_{13}$<br>$+ K_2 B_{18} B_{23} + K_\theta B_{28} B_{23}$<br>$+ K_{x\theta} B_{38} B_{33} + D_x B_{48} B_{43}$<br>$+ D_2 B_{58} B_{43} + D_2 B_{48} B_{53}$<br>$+ D_\theta B_{58} B_{53} + D_{x\theta} B_{68} B_{63}$ |
| $K_x B_{15} B_{14} + K_2 B_{25} B_{14}$<br>$+ K_2 B_{15} B_{24} + K_\theta B_{25} B_{24}$<br>$+ K_{x\theta} B_{35} B_{34} + D_x B_{45} B_{44}$<br>$+ D_2 B_{55} B_{44} + D_2 B_{45} B_{54}$<br>$+ D_\theta B_{55} B_{54} + D_{x\theta} B_{65} B_{64}$ | $K_x B_{16} B_{14} + K_2 B_{26} B_{14}$<br>$+ K_2 B_{16} B_{24} + K_\theta B_{26} B_{24}$<br>$+ K_{x\theta} B_{36} B_{34} + D_x B_{46} B_{44}$<br>$+ D_2 B_{56} B_{44} + D_2 B_{46} B_{54}$<br>$+ D_\theta B_{56} B_{54} + D_{x\theta} B_{66} B_{64}$ | $K_x B_{17} B_{14} + K_2 B_{27} B_{14}$<br>$+ K_2 B_{17} B_{24} + K_\theta B_{27} B_{24}$<br>$+ K_{x\theta} B_{37} B_{34} + D_x B_{47} B_{44}$<br>$+ D_2 B_{57} B_{44} + D_2 B_{47} B_{54}$<br>$+ D_\theta B_{57} B_{54} + D_{x\theta} B_{67} B_{64}$ | $K_x B_{18} B_{14} + K_2 B_{28} B_{14}$<br>$+ K_2 B_{18} B_{24} + K_\theta B_{28} B_{24}$<br>$+ K_{x\theta} B_{38} B_{34} + D_x B_{48} B_{44}$<br>$+ D_2 B_{58} B_{44} + D_2 B_{48} B_{54}$<br>$+ D_\theta B_{58} B_{54} + D_{x\theta} B_{68} B_{64}$ |

Stiffness matrix $[S]_{mn} = \dfrac{\alpha}{2} \displaystyle\int_0^b [\bar{S}]_{mn}\, r\, dz$

$$[\bar{S}]_{mn} = \begin{bmatrix} [\bar{S}_{11}]_{mn} & [\bar{S}_{12}]_{mn} \\ [\bar{S}_{21}]_{mn} & [\bar{S}_{22}]_{mn} \end{bmatrix}$$

TABLE 4.2 (*cont.*)

$$[\bar{S}_{22}]_{mn} =$$

$$
\begin{bmatrix}
\begin{aligned}&K_x B_{15}^2 + 2K_2 B_{25}B_{15}\\&+ K_\theta B_{25}^2 + K_{x\theta}B_{35}^2\\&+ D_x B_{45}^2 + 2D_2 B_{55}B_{45}\\&+ D_\theta B_{65}^2\end{aligned}
&
\begin{aligned}&K_x B_{16}B_{15} + K_2 B_{26}B_{15}\\&+ K_2 B_{16}B_{25} + K_\theta B_{26}B_{25}\\&+ K_{x\theta}B_{36}B_{35} + D_x B_{46}B_{45}\\&+ D_2 B_{56}B_{45} + D_2 B_{46}B_{55}\\&+ D_\theta B_{56}B_{55} + D_{x\theta}B_{66}B_{65}\end{aligned}
&
\begin{aligned}&K_x B_{17}B_{15} + K_2 B_{27}B_{15}\\&+ K_2 B_{17}B_{25} + K_\theta B_{27}B_{25}\\&+ K_{x\theta}B_{37}B_{35} + D_x B_{47}B_{45}\\&+ D_2 B_{57}B_{45} + D_2 B_{47}B_{55}\\&+ D_\theta B_{57}B_{55} + D_{x\theta}B_{67}B_{65}\end{aligned}
&
\begin{aligned}&K_x B_{18}B_{15} + K_2 B_{28}B_{15}\\&+ K_2 B_{18}B_{25} + K_\theta B_{28}B_{25}\\&+ K_{x\theta}B_{38}B_{35} + D_x B_{48}B_{45}\\&+ D_2 B_{68}B_{45} + D_2 B_{43}B_{55}\\&+ D_\theta B_{58}B_{55} + D_{x\theta}B_{68}B_{65}\end{aligned}
\\[1em]
\begin{aligned}&K_x B_{15}B_{16} + K_2 B_{25}B_{16}\\&+ K_2 B_{15}B_{26} + K_\theta B_{25}B_{26}\\&+ K_{x\theta}B_{35}B_{36} + D_x B_{45}B_{46}\\&+ D_2 B_{55}B_{46} + D_2 B_{45}B_{56}\\&+ D_\theta B_{55}B_{56} + D_{x\theta}B_{65}B_{66}\end{aligned}
&
\begin{aligned}&K_x B_{16}^2 + 2K_2 B_{26}B_{16}\\&+ K_\theta B_{26}^2 + K_{x\theta}B_{36}^2\\&+ D_x B_{46}^2 + 2D_2 B_{56}B_{46}\\&+ D_\theta B_{66}^2\end{aligned}
&
\begin{aligned}&K_x B_{16}B_{17} + K_2 B_{26}B_{17}\\&+ K_2 B_{16}B_{27} + K_\theta B_{26}B_{27}\\&+ K_{x\theta}B_{36}B_{37} + D_x B_{46}B_{47}\\&+ D_2 B_{56}B_{47} + D_2 B_{46}B_{57}\\&+ D_\theta B_{56}B_{57} + D_{x\theta}B_{66} + B_{67}\end{aligned}
&
\begin{aligned}&K_x B_{18}B_{16} + K_2 B_{28}B_{16}\\&+ K_2 B_{18}B_{26} + K_\theta B_{28}B_{26}\\&+ K_{x\theta}B_{38}B_{36} + D_x B_{48}B_{46}\\&+ D_2 B_{58}B_{46} + D_2 B_{48}B_{56}\\&+ D_\theta B_{58}B_{56} + D_{x\theta}B_{68}B_{66}\end{aligned}
\\[1em]
\begin{aligned}&K_x B_{15}B_{17} + K_2 B_{25}B_{17}\\&+ K_2 B_{15}B_{27} + K_\theta B_{25}B_{27}\\&+ K_{x\theta}B_{35}B_{37} + D_x B_{45}B_{47}\\&+ D_2 B_{55}B_{47} + D_2 B_{45}B_{57}\\&+ D_\theta B_{55}B_{57} + D_{x\theta}B_{65}B_{67}\end{aligned}
&
\begin{aligned}&K_x B_{16}B_{17} + K_2 B_{26}B_{17}\\&+ K_2 B_{16}B_{27} + K_\theta B_{26}B_{27}\\&+ K_{x\theta}B_{36}B_{37} + D_x B_{46}B_{47}\\&+ D_2 B_{56}B_{47} + D_2 B_{46}B_{57}\\&+ D_\theta B_{56}B_{57} + B_{67}\end{aligned}
&
\begin{aligned}&K_x B_{17}^2 + 2K_2 B_{27}B_{17}\\&+ K_\theta B_{27}^2 + K_{x\theta}B_{37}^2\\&+ D_x B_{47}^2 + 2D_2 B_{57}B_{47}\\&+ D_\theta B_{67}^2\end{aligned}
&
\begin{aligned}&K_x B_{17}B_{18} + K_2 B_{27}B_{18}\\&+ K_2 B_{17}B_{28} + K_\theta B_{27}B_{28}\\&+ K_{x\theta}B_{37}B_{38} + D_x B_{47}B_{48}\\&+ D_2 B_{57}B_{48} + D_2 B_{47}B_{58}\\&+ D_\theta B_{57}B_{58} + D_{x\theta}B_{67}B_{68}\end{aligned}
\\[1em]
\begin{aligned}&K_x B_{15}B_{18} + K_2 B_{25}B_{18}\\&+ K_2 B_{15}B_{28} + K_\theta B_{25}B_{28}\\&+ K_{x\theta}B_{35}B_{38} + D_x B_{45}B_{48}\\&+ D_2 B_{55}B_{48} + D_2 B_{45}B_{58}\\&+ D_\theta B_{55}B_{58} + D_{x\theta}B_{65}B_{68}\end{aligned}
&
\begin{aligned}&K_x B_{16}B_{18} + K_2 B_{26}B_{18}\\&+ K_2 B_{16}B_{28} + K_\theta B_{26}B_{28}\\&+ K_{x\theta}B_{36}B_{38} + D_x B_{46}B_{48}\\&+ D_2 B_{56}B_{48} + D_2 B_{46}B_{58}\\&+ D_\theta B_{56}B_{58} + D_{x\theta}B_{66}B_{68}\end{aligned}
&
\begin{aligned}&K_x B_{17}B_{18} + K_2 B_{27}B_{18}\\&+ K_2 B_{17}B_{28} + K_\theta B_{27}B_{28}\\&+ K_{x\theta}B_{37}B_{38} + D_x B_{47}B_{48}\\&+ D_2 B_{57}B_{48} + D_2 B_{47}B_{58}\\&+ D_\theta B_{57}B_{58} + D_{x\theta}B_{67}B_{68}\end{aligned}
&
\begin{aligned}&K_x B_{18}^2 + 2K_2 B_{28}B_{18}\\&+ K_\theta B_{28}^2 + K_{x\theta}B_{38}^2\\&+ D_x B_{48}^2 + 2D_2 B_{58}B_{48}\\&+ D_\theta B_{68}^2\end{aligned}
\end{bmatrix}
$$

in which

$$K_x = \frac{E_x t}{1 - \nu_x \nu_\theta}, \quad K_2 = \nu_\theta K_x, \quad K_{x\theta} = G_{x\theta} t, \quad K_\theta = \frac{E_\theta t}{1 - \nu_x \nu_\theta},$$

$$D_x = \frac{E_x t^3}{12(1 - \nu_x \nu_\theta)}, \quad D_2 = \nu_\theta D_x, \quad D_{x\theta} = \frac{G_{x\theta} t^3}{12}, \quad D_\theta = \frac{E_\theta t^3}{12(1 - \nu_x \nu_\theta)}.$$

The stiffness matrix is obtained in the usual way through the application of (1.30) but with the volume integral changed to an area integral. The strain matrix $[B]_m$ and the stiffness matrix $[S]_{mm}$ are listed in Tables 4.1 and 4.2 respectively. The terms $B_{ij}$ in Table 4.2 refer to the corresponding coefficients of the matrix $[B]_m$, with $ij$ indicating the location of a coefficient.

## 4.3. TRANSFORMATION OF COORDINATES

The stiffness matrices for various types of strips have so far been derived in terms of a set of local axes, two of which usually coincide with the mid-surface of a strip. Suchsti ffness matrices can be used directly in plate bending or plane stress problems because all the strips are coplanar. In folded plate structures, however, any two plates will in general meet at an angle, and in order to establish the equilibrium of nodal forces at

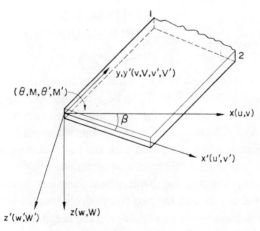

FIG. 4.3. Individual and common coordinate systems.

nodal lines common to non-coplanar strips, a common coordinate system is obviously required.

In Fig. 4.3 the individual coordinates of a strip are labelled as $x'$, $y'$, $z'$ and the common coordinates as $x$, $y$, $z$. $y$ and $y'$ are coincident with each other and also with the intersection line of two adjoining strips. The transformation of forces and displacements between the two sets of coordinate systems is given by

$$\{F\}_m = [R] \ \{F'\}_m, \tag{4.6}$$

$$\{\delta'\}_m = [R]^T \ \{\delta\}_m, \tag{4.7}$$

in which $[R]$ is the transformation matrix

$$[R] = \begin{bmatrix} [r] & [0] \\ [0] & [r] \end{bmatrix} \tag{4.8}$$

$$\text{with } [r] = \begin{bmatrix} \cos\beta & 0 & -\sin\beta & 0 \\ 0 & 1 & 0 & 0 \\ \sin\beta & 0 & \cos\beta & 0 \\ 0 & 0 & 0 & 1 \end{bmatrix} \tag{4.9}$$

$[0]$ = null matrix and $\beta$ = angle between the $x$ and $x'$ axes (clockwise positive).

The final result of the transformation is

$$\begin{aligned} \{F\}_m &= [R] \{F'\}_m \\ &= [R] [S']_{mm} \{\delta'\}_m \\ &= [R] [S']_{mm} [R]^T \{\delta\}_m \\ &= [S]_{mm} \{\delta\}_m. \end{aligned} \tag{4.10}$$

Once the stiffness matrix of a strip has been transformed into the common coordinate system, it is ready to be assembled into the overall stiffness matrix of the structure in the conventional manner.

In the process of transformation, two points stand out in favour of the finite strip method when compared with the finite element method. The first concerns the compatibility of displacements after transformation. In the finite strip method, the displacement functions have been chosen in such a way that $u$ and $w$, the two components that are involved in the transformation, have the same variation in the longitudinal ($y$) direction, and a rotation of coordinate axes $x$ and $z$ will consequently not affect the compatibility of displacements at the nodal lines. This is, however, not so for flat shell finite elements since the majority of such elements use different order polynomials for the $u$ and $w$ displacement components. Even if both $u$ and $w$ are compatible displacement functions when examined individually, after transformation they will be combined in a certain proportion depending on the direction cosines of the element concerned and compatibility of displacements will in general be lost. Secondly, for most flat shell finite elements, only five DOF $(u', v', w', \theta'_x, \theta'_y)$ exist at each node, and complications usually arise when they are transformed into the six DOF $(u, v, w, \theta_x, \theta_y, \theta_z)$ common coordinate system. As a result, in finite element analysis either an artifical sixth DOF[16] or, indeed, some artificial stiffness coefficients[17] have to be inserted, or else a set of surface coordinate axes,[18] the orientation of which changes from node to node, has to be used in order to retain the five-DOF system. Meanwhile the strips always retain their four DOF per nodal line, and only the standard transformation used in plane frame analysis is needed.

## 4.4. APPLICATIONS TO FOLDED PLATE STRUCTURES

### 4.4.1. PRISMATIC FOLDED PLATE UNDER RIDGE LOADS[19]

A typical folded plate cross-section (Fig. 4.4) was analysed by the finite strip method and compared with the results obtained by De Fries-Skene and Scordelis.[3] Four different span lengths of 100 m, 70 m, 30 m, and 10 m were used in order to demonstrate the applicability of the method to long and short shells. Poisson's ratio was taken as zero in the examples.

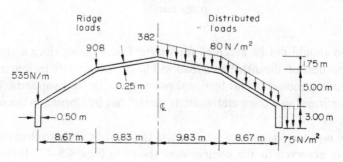

FIG. 4.4. Folded plate sectional dimensions and loadings.

The structures are assumed to be under ridge loads, although the amount of work involved in analysing problems with distributed loads would be just the same. Note that the ridge loads are really equivalent loads obtained from conventional elementary (one-way slab) analysis, and they were used in the example of the Task Force report.[1] In practice, such a simpli-

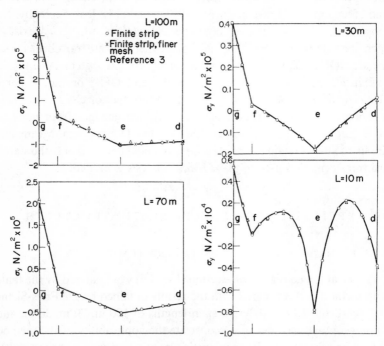

FIG. 4.5. Longitudinal stress $\sigma_y$ at mid-span for various span lengths $L$ (ridge loads).

fication should not be accepted for shorter folded plates, since a significant portion of the distributed load will be carried directly by longitudinal bending moments and torsional moments to the end supports. This was confirmed in a comparative study carried out by Cheung in the same paper.

Five non-zero harmonics were used in the analysis, and excellent agreement is observed in the comparisons shown in Figs. 4.5–4.7. It should be pointed out that since the transverse membrane force $P$ is constant

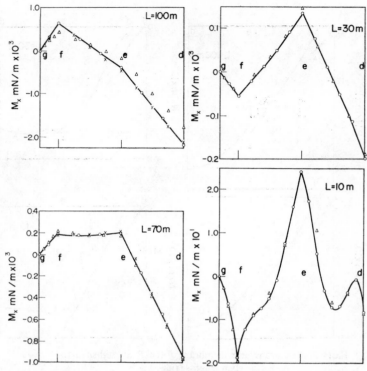

FIG. 4.6. Transverse moment $M$ at mid-span for various span lengths $L$
(ridge loads).

across the width of a lower order strip (due to the assumption that $u$
is linear), the values of $P$ were plotted at the middle of each strip to avoid a
stepwise presentation.

### 4.4.2. CURVED BEAM PROBLEM[14]

The accuracy of the curved finite strip is tested in this example in which
a curved beam is analysed by ordinary curved beam theory neglecting
warping torsion and by the finite strip method, for the four cross-sections
depicted in Fig. 4.8a. By maintaining a constant value of the length of
the beam $R\alpha$ and denoting the ratio between the bending moments in a
curved beam and in the corresponding straight beam of equal length
$L = R\alpha$ by

$$\varrho = \frac{M_{\text{curved}}}{M_{\text{straight}}},$$

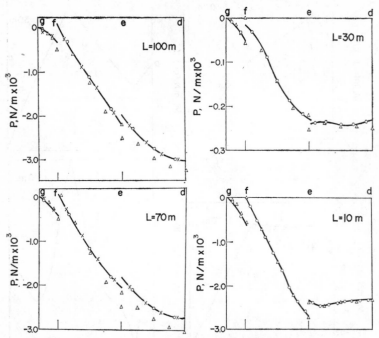

FIG: 4.7. Plate transverse membrane force $P$ at mid-span for various span lengths (ridge loads).

it is possible to plot $\varrho$ against the included angle $\alpha$. Figure 4.8b and c demonstrate clearly the good agreement between the finite strip and curved beam theory results. In Fig. 4.8c the moments are plotted 1 m away from the concentrated mid-span load because too many Fourier series terms would be needed to represent accurately the value of the bending moment directly underneath or in the immediate vicinity of the concentrated load.

## 4.5. APPLICATION TO BOX GIRDER BRIDGES

### 4.5.1. INTRODUCTION

Certain special features not commonly associated with folded plate roof structures exist in the analysis of box girder bridges. These features will be discussed in conjunction with finite strip analysis.

(i) Due to the fact that the primary function of a box girder bridge is to carry traffic, the top flanges of a box girder bridge is acted upon by groups of point loads or patch loads representing the wheel loadings of trucks and other motor vehicles.

At the immediate vicinity of a point load, the stresses and moments obtained by the summation of a Fourier series tend to converge very slowly, and sometimes as many as fifty non-zero harmonics have to be taken before a "converged" solution is arrived at. On the other hand, for stresses and moments at points a small distance away from the point load, hardly any change will be observed after the first fifteen or so non-zero harmonics. Since the stresses and moments under a point load is mathematically infinite, it is probably not worth while to spend any significant amount of computer time to try to achieve "convergence" at such location.

In any case, the reinforcements in the bridge are designed to resist the total integrated force over some part of the cross-section, and not according to the exact stress distribution pattern.

As mentioned above, many terms of the series are required for bridge analysis due to the presence of concentrated loads; hence, boundary conditions other than simple supports at the two ends cannot be included because the terms of the series will then be coupled. Such coupling tends to create a stiffness matrix of very large bandwidth for concentrated load cases, and is therefore uneconomical.

(ii) Box girder bridges of longer spans very often are supported by intermediate discrete columns. Such columns are sometimes located at odd positions which are determined by site conditions or by the roadways passing underneath. Diaphragms are normally provided at these intermediate support sections to ensure stability and to reduce stress concentration effects.

The method of solution for column-supported slab bridges discussed in Section 2.7.2 can be applied directly to the present situation.

(iii) Internal diaphragms very often are present, especially for bridges of steel construction, in which stability of the various plate elements has emerged to be one of the critical factors in design. For concrete bridges, however, recent research work[19, 20] has demonstrated that the boxes themselves possess sufficient torsional rigidity for adequate transverse load distribution, and therefore only end diaphragms or diaphragms over intermediate supports are regarded as absolutely essential. This is of some importance because the construction of concrete diaphragms can be a costly and time-consuming process, and their number should be limited to a minimum.

The analysis of box girder bridges with intermediate rigid diaphragms was carried out by Scordelis *et al.*,[21] using the elasticity method in conjunction with a flexibility approach. This procedure can be adapted to finite strip analysis and will therefore be presented in detail.

Before proceeding any further it would be opportune to sum up the main points of the elasticity method and to define some of the terms that will appear in the following discussions.

(a) *Definitions*

(1) Internal diaphragms may be "movable" or "immovable". Immovable ones are connected to an intermediate internal support that prevents displacements in the plane of the diaphragm, while "movable" ones may move as a rigid body within their own plane since they are connected to the box girder only. For immovable diaphragms, the general procedure is similar to the one discussed in Section 2.7.2 and will not be discussed here.

(2) A box girder that has two end diaphragms but no intermediate diaphragms is referred to as a primary structure.

(b) *Analysis of primary structure*

The basic structural element in this analysis is a single plate having a width equal to the distance between two consecutive longitudinal joints (e.g. top or bottom flange or web of a cell in a box girder bridge) and a length equal to the overall length of the bridge. (A Fourier expansion in the longitudinal direction is used.) A stiffness matrix relates the edge forces (four per edge) and the edge displacements (also four per edge).

Equilibrium equations are established at the joints. The joint loadings correspond to (1) line load or concentrated load applied directly to the joints, or (2) fixed-edge loads equal but opposite to the reactions caused by loads acting on a plate which is clamped along the longitudinal edges.

The whole process is thus similar to a plane frame analysis using the displacement method.

(c) *Analysis of a bridge with intermediate diaphragms*

Structures with intermediate diaphragms are analysed by solving the

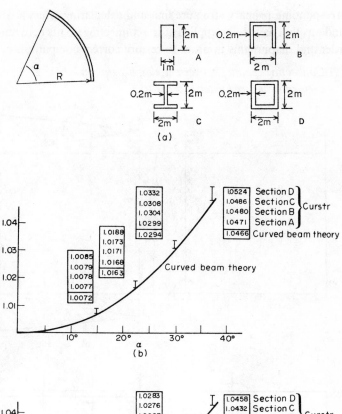

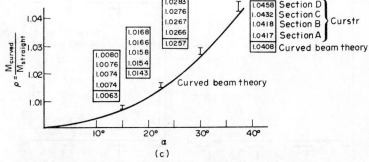

FIG. 4.8. Example 1-Curved beam problem (Meyer and Scordelis [14]). (a) Curved beam dimensions and cross-section types. (b) Moment variation for uniform load. (c) Moment variation for concentrated load. (Original units in FPS.)

corresponding primary structure first and calculating afterwards the redundants that have to be applied at the connecting points between the box
girder and diaphragms in order to restore correct compatibility.

*1. Connecting points between a diaphragm and box girder and the
corresponding redundant forces*

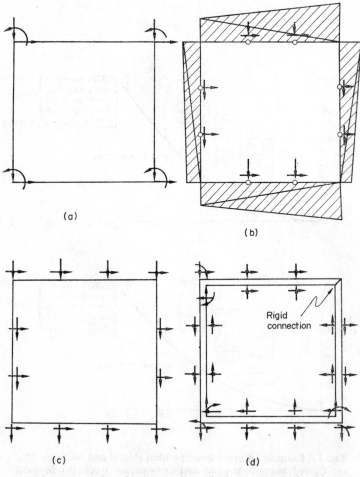

FIG. 4.9. Box girder with intermediate diaphragm. (a) Redundants at
joints. (b) Redundants at middle third points of plate. (c) Redundants $\{X\}$
acting on box girder system. (d) Redundants $\{X\}$ acting on box girder
"supported" diaphragm system.

A diaphragm is connected with the box girder all along its edges and, strictly speaking, there is an infinite number of connecting points. In actual computation, the problem has to be reduced to a manageable size, and Scordelis et al.[21] assumed that a diaphragm is connected to each cell of the box girder at the joints and also at the middle third points of each plate. In a recent report the effect of a rigid diaphragm is simulated by Meyer[22] using restraints at cell corners only.

It was mentioned in (b) that four edge forces exist at a longitudinal joint. Among these the longitudinal shear force cannot be used as a redundant because the diaphragm is assumed to be perfectly flexible out of plane. As a result the redundant forces at each longitudinal joint consist of a vertical force, a horizontal force, and a bending moment in the plane of the diaphragm (Fig 4.9a).

The redundants for the middle third points consist of distributed normal and tangential forces having triangular variations between the two longitudinal edges of the plate (Fig. 4.9b). Moment loads are not used. All of the redundant forces are assumed to be uniformly distributed over the thickness of the diaphragm.

*2. Redundant force vector $\{X\}$ and absolute displacements $\{\delta\}_x$ at the connecting points (Fig. 4.9c)*

Consider the box girder and diaphragm separately. The self-equilibrating redundant forces $\{X\}$ acting on the box girder will cause absolute displacements $\{\delta_x\}$ at all the connecting points.

*3. Redundant force vector $\{\overline{X}\}$ and relative displacement $\{\overline{\delta}\}_x$ (Fig. 4.9d)*

Assume that the rigid diaphragm is now attached to the box girder in any statically determinate manner (e.g. rigidly connected at corner $C$ in Fig. 4.9d). The equal and opposite redundant forces $\{\overline{X}\}$ acting on the structure shown in Fig. 4.9d consisting of the box girder and diaphragm will cause relative displacements $\{\overline{\delta}\}_x$. Note that $\{\overline{X}\} = \{X\} - 3$ because one of the joints is now used as a support and three equilibrium equations can now be established.

*4. Relationship between $\{X\}$ and $\{\overline{X}\}$*

The two vectors $\{X\}$ and $\{\overline{X}\}$ are related to each other through

$$\{X\} = [B]\{\overline{X}\}. \tag{4.11}$$

By comparing Fig. 4.8c it is seen that $\{X\}$ and $\{\bar{X}\}$ are identical at all points except for $C$, where the three components of $\{X\}$ are linear combinations of $\{\bar{X}\}$, or to be more specific the reactions at $C$ due to $\{\bar{X}\}$.

## 5. Relationship between displacements of the absolute and relative systems

The matrix $[B]$ relates all the redundant forces acting on the box girder system to the redundant forces acting on the box girder and "supported" diaphragm system. It can be shown through virtual work[†] that the transpose of the $[B]$ matrix will also relate the relative displacements $\{\delta\}$ of the box girder–"supported" diaphragm system to the absolute displacements $\{\delta\}$ of the box girder system.

Thus for redundant forces only

$$\{\delta\}_x = [B]^T \{\delta\}_x \tag{4.12a}$$

and for external loadings only

$$\{\delta\}_0 = [B]^T \{\delta\}_0. \tag{4.12b}$$

## 6. Compatibility conditions

From the flexibility analysis of the primary structure,

$$\{\delta\}_x = [F] \{X\}, \tag{4.13a}$$

substituting (4.11) and (4.13a) into (4.12a),

$$\{\delta\}_x = [B]^T [F] \{X\}$$
$$= [B]^T [F] [B] \{\bar{X}\} = [\bar{F}] \{\bar{X}\}. \tag{4.13b}$$

---

† The work done for the two systems should be identical, i.e.

$$\{\delta\}_x^T\{X\} = \{\delta\}_x^T\{\bar{X}\} \quad \text{or} \quad \{\delta\}_x^T[B]\{\bar{X}\} = \{\delta\}_x^T\{\bar{X}\}$$

Transposing,

$$\{\bar{X}\}^T[B]^T\{\delta\}_x = \{\bar{X}\}^T\{\delta\}_x.$$

Since the relationship is valid for any set of redundant forces, $\{\bar{X}\}^T$ can be cancelled off and we finally obtain

$$\{\delta\}_x = [B]^T\{\delta\}_x.$$

The redundants $\{\overline{X}\}$ may be computed by setting the relative displacements between the "supported" diaphragm and the box girder equal to zero:

$$\{\delta\} = \{\delta\}_0 + \{\delta\}_x = 0 \tag{4.14}$$

or

$$\{\delta\}_0 + [\overline{F}]\{\overline{X}\} = 0.$$

The redundant forces $\{X\}$ acting on the box girder system may be determined through (4.11).

A computer program STRIP[25] for the analysis of straight and curved box girder bridges based on the finite strip method now forms a part of the bridge design computer programs issued by the Department of the Environment of Great Britain.

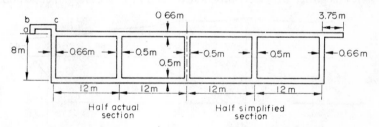

FIG. 4.10. Cross-sectional dimensions of box girder bridge.

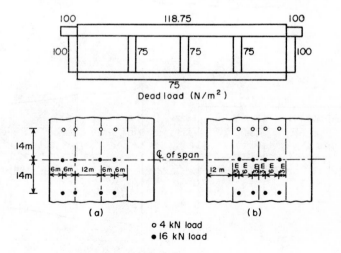

FIG. 4.11. Schemes of loadings.

### 4.5.2. RIGHT BOX GIRDER BRIDGE BY LOWER ORDER STRIP[9]

A typical box girder bridge (Fig. 4.10) which was analysed by Chu and Dudnik[5] using the elasticity method is taken as an example to show the accuracy of the lower order strip analysis. In order to compare the results with those given in reference 5, the sidewalk portion is also similarly replaced by a cantilever slab 0.66 m thick and 3.75 m wide, although no more effort or computer time is required if the actual cross-section were used in calculation.

Three types of loadings were applied to the bridge deck and their magnitudes and distribution are given in Fig. 4.11. They are respectively the dead load of the structure and two types of truck loadings.

The box girder is divided into 93 strips and there are altogether 90 nodal

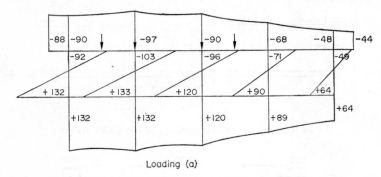

Loading (a)

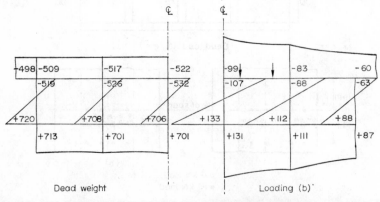

Dead weight               Loading (b)

FIG. 4.12. Longitudinal stresses $\sigma_y$ (in $N/m^2 \times 144^{-1}$).

lines. The top slab has to be divided into a much finer mesh because of the concentrated wheel loadings.

Eleven terms of the Fourier series have been used for the analysis. This would be on the low side for representing the effect of a single point load; fortunately, the severity of stress concentration is somewhat alleviated when a group of point loads near to each other (such as truck wheel loadings) is present. For dead loads which are symmetrical with respect to the centre of the span, only the odd terms of the series need to be considered since the even terms produce zero value answers.

The nodes have been numbered to produce a matrix with the narrowest half-bandwidth. For the present bridge analysis the execution time on the medium speed multi-programming IBM 360–50 computer is 25 min. The above solution time can actually be reduced rather drastically because subsequent experience shows that only one strip is needed in order to represent adequately the web of the bridge, whereas three strips have actually been used for the present solution.

The transverse and longitudinal stresses are positive when tensile, while the bending moments $M_x$ and $M_y$ referred to the individual coordin-

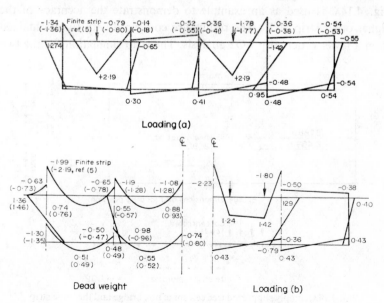

Fig. 4.13. Transverse moments $M_x$ (in kN-m/m).

ate system of the strips follow the usual sign convention given in Timo-shenko and Woinowsky-Kreiger[23]. Here the $x$ axes always point towards the right for horizontal strips and downwards for vertical strips.

The longitudinal stresses $\sigma_y$ and the transverse bending moments $M_x$ at the centre of the span for all three loading cases are presented in Figs. 4.12 and 4.13, and are compared with those given by reference 5. In general, the values of longitudinal stresses and transverse moments due to dead load are slightly below those given by reference 5, although the pattern of variation is the same in both cases. For the finite strip analysis the maximum beam moment due to dead weight checked closely with the sum of the resisting moments corresponding to the longitudinal stresses multiplied by their respective moment arms, and this computation provides for an additional check on the accuracy of the solution.

### 4.5.3. RIGHT BOX GIRDER BRIDGE BY HIGHER ORDER STRIP[13] (HO3)

The three-cell spine box bridge under sixteen wheel loads, shown in Fig. 4.14A, is used as an example to demonstrate the accuracy of the higher order strip HO3. The results are compared with those obtained from a lower order strip LO2 analysis. The strip simulation for the two

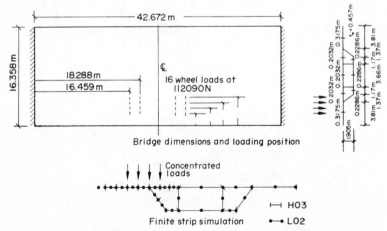

FIG. 4.14A. Simply supported tree cell spine box bridge and the finite strip simulation (Loo[24]).

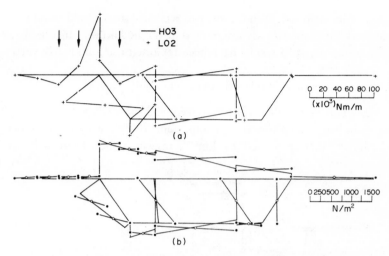

FIG. 4.14B. (a) Transverse moment, and (b) in-plane stress profiles for the three cell spine box bridge under sixteen wheel loads (at section under the second axle) (Loo[24]).

analyses is given in Fig. 4.14A. Twenty-five harmonics are used in both analyses. The transverse bending moment and the transverse in-plane stress profiles at the section under the second axle are plotted in Fig. 4.14B, a and b respectively. It can be seen that good agreement between the two cases is obtained. Since the transverse strain across the width of a

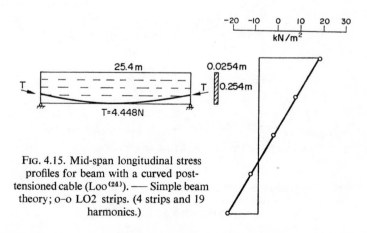

FIG. 4.15. Mid-span longitudinal stress profiles for beam with a curved post-tensioned cable (Loo[24]). —— Simple beam theory; o–o LO2 strips. (4 strips and 19 harmonics.)

lower order strip is constant, a stepped representation would result if the transverse in-plane stresses are plotted at the nodal lines. In Fig. 4.13b the lower order strip results have been replotted using mid-strip values.

### 4.5.4. ANALYSIS OF PRESTRESSED FORCES (LO2)[24]

Many concrete bridges are prestressed, and the analysis of prestressed forces in bridge structure is very important. By following the treatment of curved cables proposed by Lin[26], in which the prestressed force

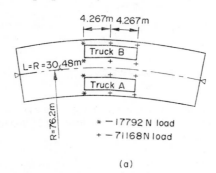

(a)

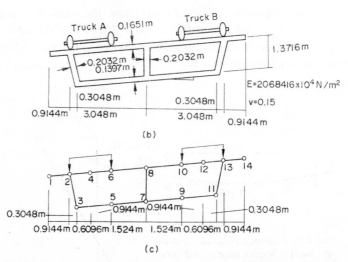

(b)

(c)

FIG. 4.16. Curved box girder bridge (Meyer and Scordelis[14]). (a) Plan view. (b) Cross-section. (c) Curved strip idealization and nodal joint numbering. (Original unit in FPS.)

system is resolved into concentrated end loads and distributed upward pressure along the span, it is possible to compute the effects of prestressing with the standard finite strip procedure.

The simply supported deep beam with a curved frictionless post-tensioned cable is shown in Fig. 4.15, and four equal strips were used to simulate the beam. The distribution of mid-span longitudinal stress is plotted in Fig. 4.15. Comparison between the results of the finite strip analysis and those due to simple beam (prestressed) theory is excellent.

### 4.5.5. CURVED BOX GIRDER BRIDGES BY LOWER ORDER STRIP[14] (LO2)

The two-cell box girder bridge shown in Fig. 4.16a subtends an angle of 22.92°, and was analysed with a standard AASHO-truck placed at two

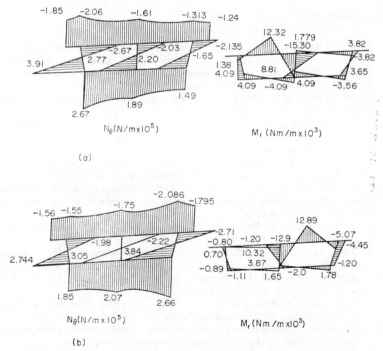

FIG. 4.17. Longitudinal stress resultants and $N_\theta$ transverse moments $M$ at mid-span of box girder bridge (Meyer and Scordelis [14]). (a) Truck position A. (b) Truck position B. (Original unit in FPS.)

different positions (Fig. 4.16b). Only the downward-loading component has been considered.

The longitudinal stress resultants $N_\theta$ and transverse bending moments $M_r$ for the two truck positions are shown in Fig. 4.17. In spite of its smaller statical moment, truck A produces higher maximum stresses and transverse bending moments than truck B. This is similar to the phenomenon exhibited by a curved plate, where the more flexible outer part tends to distribute the load to other parts of the bridge, while the stiffer inner part tends to carry the load alone, with very little load distribution taking place.

Using fifty terms of the Fourier series, the total execution time on the CDC 6400 for each run (one loading case) is 27.2 sec.

## REFERENCES

1. Report of the Task Committee on Folded Plate Construction, Phase I report on folded plate construction, *Am. Soc. Civ. Engrs* **89**, ST, 365–405 (December 1963).
2. J. E. GOLDBERG and H. L. LEVE, *Theory of Prismatic Folded Plate Structures*, IABSE Publications, Zurich, Switzerland, vol. **17**, pp. 59–86 (1957).
3. A. DE FRIES-SKENE and A. C. SCORDELIS, Direct stiffness solution for folded plates, *Am. Soc. Civ. Engrs* **90**, ST, 15–47 (August 1964.).
4. A. C. SCORDELIS, *Analysis of Simply Supported Box Girder Bridges*, Structural Engineering and Structural Mechanics Report No. SESM 66–17, Univ. of California, Berkeley (October 1966).
5. K. H. CHU and E. DUDNIK, *Concrete Box Girder Bridges Analysed as Folded Plates*, Concrete Bridge Design, ACI Publications SP-23, 221–246, 1969.
6. J. E. GOLDBERG, M. CASTILLO, and R. K. KEPLAR, An exact theory of horizontal curved girder structures, *Proceedings IASS Symposium on Folded Plates and Prismatic Structures, Vienna*, 1970.
7. K. H. CHU and S. G. PINJARKAR, Analysis of horizontally curved box girder bridges, *Am. Soc. Civ. Engrs* **97**, ST. 10, 2481–2501 (October 1971).
8. M. REHN and G. SVED, Stiffness matrices of sector shaped thin shell elements, *Bulletin International Association for Shell Structures*, No. 45, 3–22 (March 1971).
9. Y. K. CHEUNG, Folded plate structures by finite strip method, *Am. Soc. Civ. Engrs* **95**, No. ST 12, 2963–79 (December 1969).
10. Y. K. CHEUNG, *Analysis of Box Girder Bridges by Finite Strip Method*, Concrete Bridge Design, ACI Publications SP-26, 357–378 (1971).
11. Y. K: CHEUNG, Analysis of orthotropic prismatic folded plates, *Proceedings IASS Symposium of Folded Plates and Prismatic Structures, Vienna, 1970.*
12. K. J. WILLIAM and A. C. SCORDELIS, Analysis of eccentrically stiffened folded plates, *Procedings IASS Symposium on Folded Plates and Prismatic Structures, Vienna 1970.*

13. Y. C. Loo and A. R. Cusens, Developments of the finite strip method in the analysis of cellular bridge decks, *Developments in Bridge Design and Construction* (ed. Rodsey *et al.*), Crosby, Lockwood (October 1971).
14. C. Meyer and A. C. Scordelis, Analysis of curved folded plate structures, *Am. Soc. Civ. Engrs* 97, No. ST 10, 2459–80 (October 1971).
15. M. S. Cheung and Y. K. Cheung, *Analysis of Curved Box Girder Bridges by Finite Strip Method*, Publications International Association for Bridges and Structural Engineering, vol. 31-I, pp. 1-20, 1971.
16. O. C. Zienkiewicz and Y. K. Cheung, Finite element method analysis of arch dam shells and comparison with finite difference prodecure, *Procedings Symposium of Theory of Arch Dams, Southampton University, 1964*, Pergamon Press, 1956.
17. O. C. Zienkiewicz, C. J. Parekh, and I. P. King, Arch dams analysed by a linear finite element shell program, *Proceedings Symposium Arch Dams, Institution Civil Engineers, London, 1968.*
18. R. W. Clough and C. P. Johnson, A finite element approximation for the analysis of thin shells, *Int. J. Solids Struct.* 4, 43–60 (1968).
19. J. C. Chapman, P. J. Dowling, P. T. K. Lim, and R. L. Billington, The structural behaviour of steel and concrete bridges, *Struct. Engr.* 49 (3), 111–120 (March 1971).
20. R. G. Sisodiya, A. Ghali, and Y. K. Cheung, Diaphragms in single- and double-cell box girder bridges with varying angle of skew, *ACI J* 415-19, 1972.
21. A. C. Scordelis, R. E. Davis, and K. S. Lo, *Load Distribution in Concrete Box Girder Bridges*, Concrete Bridge Design, ACI Publications SP-23, 117-136, 1969.
22. C. Meyer, *Analysis and Design of Curved Box Girder Bridges*, Structural Engineering and Structural Mechanics Report, No. SESM 70-22, Univ. California, Berkeley, 1970.
23. S. Timoshenko and S. Woinowsky-Kreiger, *Theory of Plates and Shells*, 2nd edn., McGraw-Hill, 1959.
24. Y. C. Loo, Developments and applications of the finite strip method in the analysis of right bridge decks, Ph.D. Thesis, University of Dundee, 1972.
25. R. Travers Morgan & Partners, *STRIP Report and User Manual*, Department of the Environment, 1972, March.
26. T. Y. Lin, *Design of Prestressed Concrete Structures*, John Wiley & Sons, 2nd edn., New York, 1963.

# CHAPTER 5

# Vibration and stability of plates and shells

## 5.1. MATRIX THEORY OF FREE VIBRATIONS

The analysis of a variety of elastic structures subject to static loads have been discussed in the previous chapters. It can be observed that all types of formulations will eventually lead to the same matrix equation of

$$[K]\{\delta\} = \{P\} \tag{5.1}$$

in which $[K]$ is the stiffness matrix of structure, $\{\delta\}$ the vector containing all nodal displacement parameters, and $\{P\}$ the vector containing all nodal forces.

If the structure is moving then it is also possible to reduce the dynamic problem to a static one by applying D'Alembert's principle of dynamic equilibrium in which an inertia force equal to the product of the mass and the acceleration is assumed to act on the structure in the direction of negative acceleration. Thus at any instant of time the equilibrium equation for a structure in which both damping and external excitation forces are assumed to be non-existent is

$$[K]\{\delta(t)\} = -[[M]^c + [M]^d]\{\ddot{\delta}(t)\} = -[M]\{\ddot{\delta}(t)\}. \tag{5.2a}$$

where $\delta(t)$ is now a function of time and $\cdot\cdot$ represents $\partial^2/\partial t^2$.

In the above equation $[M]^c$ is a diagonal matrix of concentrated or line masses at the nodal lines, and is simply equal to zero when no such concentrated or line masses are acting on the structure, and $[M]^d$ is an overall mass matrix of the structure assembled from individual element consistent mass matrices[†] $[M]^e$. The assembly process for mass matrices and for stiffness matrices are identical and will be discussed in Chapter 8.

For free vibration, the system is vibrating in a normal mode, and it is possible to make the substitutions

$$\{\delta(t)\} = \{\delta\} \sin \omega t, \quad \{\ddot{\delta}(t)\} = -\omega^2\{\delta\} \sin \omega t \tag{5.2b}$$

into (5.2) to obtain

$$([K] - \omega^2[M]) \{\delta\} = 0, \tag{5.3a}$$

where $\omega$ is the natural frequencies of the modes and the common term $\sin \omega t$ has been cancelled out.

It is possible to transform (5.3a) into

$$[K]^{-1} [M] \{\delta\} = \frac{1}{\omega^2} \{\delta\} \tag{5.3b}$$

which becomes, therefore, a standard eigenvalue problem with the form

$$[A] \{x\} = \lambda\{x\}. \tag{5.3c}$$

Various schemes have been developed for solving eigenvalue equations, and some of the more standard techniques are described in texts such as the one by Bishop et al.[4] For more advanced techniques the mass condensation method proposed by Irons[5] and the subspace iteration method developed by Bathe[6] should be used.

Note that a direct solution (5.3b) is uneconomical because although both $[K]^{-1}$ and $[M]$ are symmetrical, the product $[K]^{-1} [M]$, however, is in general not symmetrical, and in practice some form of transformation similar to the process to be described in Chapter 8 should be applied first.

All the examples in this chapter have been solved by a eigenvalue program developed by Anderson.[7].

---

† The "term consistent mass matrix" is used for distributed masses and the matrix is derived through virtual work or the minimum total potential energy principle. Such a matrix is usually a full matrix and is different from the previously prevalent lumped mass matrix which is a diagonal matrix resulting from arbitrary lumping of distributed masses at the nodal lines. The fact that the latter procedure will lead to a poorer approximation was simultaneously recognized by Archer[1] and by Leckie and Lindberg[2], and a general presentation in the context of finite element analysis was presented by Zienkiewicz and Cheung.[3]

## 5.2. DERIVATION OF CONSISTENT MASS MATRIX OF A STRIP

The displacement function for any strip has the general form of

$$\{f\} = [N]\{\delta\} = \sum_{m=1}^{r} [N]_m \{\delta\}_m \tag{1.16b}$$

in which both $\{f\}$ and $\{\delta\}$ are time dependent.

If the mass is distributed throughout the strip, any acceleration will give rise to distributed inertia forces of the magnitude

$$\{q\} = -\varrho t \frac{d^2\{f\}}{dt^2} \tag{5.4a}$$

with $\varrho$ being the mass per unit volume and $t$ the thickness of strip.

Substituting (1.16b) into (5.4a) and taking note of (5.2b),

$$\{q\} = -\varrho t [N]\{\ddot{\delta}(t)\}$$
$$= -\varrho t \omega^2 [N]\{\delta\} \sin \omega t. \tag{5.4b}$$

From (5.3a) it is seen that the term $\sin \omega t$ will be cancelled out and therefore duly dropped in all subsequent derivations. Equation (5.4b) is now simply

$$\{q\} = \varrho t \omega^2 [N]\{\delta\}. \tag{5.4c}$$

Based on (1.33) it is possible at this stage to obtain equivalent nodal forces for the distributed inertia loadings. Thus

$$F = \int [N]^T \{q\}\, d\,(\text{area})$$

$$= \int \varrho t \omega^2 [N]^T [N]\{\delta\}\, d\,(\text{area})$$

$$= \int \varrho t \omega^2 [[N]_1 [N]_2 [N]_3 \ldots [N]_r]^T [[N]_1 [N]_2 [N]_3 \ldots [N]_r]\{\delta\}\, d\,(\text{area})$$

$$= \int \varrho t \omega^2 \begin{bmatrix} [N]_1^T [N]_1 & [N]_1^T [N]_2 & [N]_1^T [N]_3 \ldots [N]_1^T [N]_r \\ [N]_2^T [N]_1 & [N]_2^T [N]_2 & [N]_2^T [N]_3 \ldots [N]_2^T [N]_r \\ [N]_3^T [N]_1 & [N]_3^T [N]_2 & [N]_3^T [N]_3 \ldots [N]_3^T [N]_r \\ \cdots & \cdots & \cdots \quad \cdots \quad \cdots \\ [N]_r^T [N]_1 & [N]_r^T [N]_2 & [N]_r^T [N]_3 \ldots [N]_r^T [N]_r \end{bmatrix} \{\delta\}\, d\,(\text{area})$$

$$= \omega^2 [M]^e \{\delta\} \tag{5.5}$$

Thus the basic unit submatrix in a consistent mass matrix is

$$[M]^e_{mn} = \int \varrho t [N]^T_m [N]_n \, d \text{ (area)}. \tag{5.6}$$

## 5.3. CONSISTENT MASS MATRICES OF PLATE STRIPS IN BENDING

### 5.3.1. RECTANGULAR BENDING STRIP (LO2)

Using the rectangular bending strip discussed in Section 2.2.1 the matrix $[N]_m$ is defined as

$$[N]_m = [(1-3\bar{x}^2+2\bar{x}^3), \quad x(1-2\bar{x}+\bar{x}^2), \quad (3\bar{x}^2-2\bar{x}^3), \quad x(\bar{x}^2-\bar{x})]Y_m$$
$$= [C]Y_m.$$

The product $\varrho t$ in (5.6) is usually assumed to be constant in the transverse direction, and we have for the mass matrix

$$[M]^e_{mn} = \int_0^b [C]^T [C] \, dx \int_0^a \varrho t Y_m Y_n \, dy. \tag{5.7}$$

The integrals with respect to $dx$ are quite simple and have been worked out by hand, and the submatrix $[M]^e_{mn}$ in a semi-complete form is given in Table 5.1.

TABLE 5.1. CONSISTENT MASS MATRIX OF A LO2 RECTANGULAR STRIP WITH ANY END CONDITIONS

$$[M]^e_{mn} = \int_0^a
\begin{bmatrix}
\dfrac{13b}{35} & & \text{Symmetrical} & \\[2ex]
\dfrac{11b^2}{210} & \dfrac{b^3}{105} & & \\[2ex]
\dfrac{9b}{70} & \dfrac{13b^2}{420} & \dfrac{13b}{35} & \\[2ex]
-\dfrac{13b^2}{420} & -\dfrac{3b^3}{420} & -\dfrac{11b^2}{210} & \dfrac{b^3}{105}
\end{bmatrix}
\varrho t(y) Y_m Y_n \, dy$$

If the product $\varrho t$ is further assumed to be constant within the strip, then for all the six basic functions listed in Section 1.3.1 the integral $\int_0^a \varrho t Y_m Y_n dy$ in (5.7) will be zero for $m \neq n$ because of their orthogonal properties. This will mean that all off-diagonal submatrices in the mass matrix will become zero, and as a result the consistent mass matrix can be greatly simplified.

For the special case of simply supported strip with constant thickness the integral in Table 5.1 is reduced to a standard form of

$$
\left.
\begin{aligned}
\int_0^a \varrho t Y_m Y_n \, dy &= \varrho t \int_0^a \sin^2 k_m y \, dy \\
&= \frac{\varrho t a}{2} && \text{for } m = n \\
&= 0 && \text{for } m \neq n.
\end{aligned}
\right\}
\tag{5.8}
$$

### 5.3.2. RECTANGULAR BENDING STRIP (HO2)

For this strip a quintic polynomial is used in the transverse direction and the displacement function is described in Section 2.2.2. The corresponding matrix $[N]_m$ is given as

$$
\begin{aligned}
[N]_m &= [(1 - 10\bar{x}^3 + 15\bar{x}^4 - 6\bar{x}^5), \quad x(1 - 6\bar{x}^2 + 8\bar{x}^3 - 3\bar{x}^4), \\
&\quad x^2(\tfrac{1}{2} - \tfrac{3}{2}\bar{x} + \tfrac{3}{2}\bar{x}^2 - \tfrac{1}{2}\bar{x}^3), \quad (10\bar{x}^3 - 15\bar{x}^4 + 6\bar{x}^5), \\
&\quad x(-4\bar{x}^2 + 7\bar{x}^3 - 3\bar{x}^4), \quad x^2(\tfrac{1}{2}\bar{x} - \bar{x}^2 + \tfrac{1}{2}\bar{x}^3)]Y_m \\
&= [C]Y_m.
\end{aligned}
$$

Once more it is possible to apply (5.7) for the computation of the mass matrix, which has also been worked out in a semi-complete form in Table 5.2.

TABLE 5.2. CONSISTENT MASS MATRIX OF A HIGHER ORDER STRIP HO2

$$[M]^e_{mn} = \int_0^a \varrho t\, Y_m Y_n\, dy$$

| | | | | | |
|---|---|---|---|---|---|
| $+\dfrac{181}{462}b$ | $+\dfrac{311}{4620}b^2$ | $+\dfrac{34931}{55440}b^3$ | $+\dfrac{25}{231}b$ | $-\dfrac{151}{4620}b^2$ | $+\dfrac{181}{55440}b^3$ |
| $+\dfrac{311}{4620}b^2$ | $+\dfrac{52}{3465}b^3$ | $+\dfrac{23}{18480}b^4$ | $+\dfrac{151}{4620}b^2$ | $-\dfrac{19}{1980}b^3$ | $+\dfrac{13}{13860}b^4$ |
| $+\dfrac{34931}{55440}b^3$ | $+\dfrac{23}{18480}b^4$ | $+\dfrac{1}{9240}b^5$ | $+\dfrac{181}{55440}b^3$ | $-\dfrac{13}{13860}b^4$ | $+\dfrac{1}{11088}b^5$ |
| $+\dfrac{25}{231}b$ | $+\dfrac{151}{4620}b^2$ | $+\dfrac{181}{55440}b^3$ | $+\dfrac{181}{462}b$ | $-\dfrac{311}{4620}b^2$ | $+\dfrac{281}{55440}b^3$ |
| $-\dfrac{151}{4620}b^2$ | $-\dfrac{19}{1980}b^3$ | $-\dfrac{13}{13860}b^4$ | $-\dfrac{311}{4620}b^2$ | $+\dfrac{11}{693}b^3$ | $-\dfrac{23}{18480}b^4$ |
| $+\dfrac{181}{55440}b^3$ | $+\dfrac{13}{13860}b^4$ | $+\dfrac{1}{11088}b^5$ | $+\dfrac{281}{55440}b^3$ | $-\dfrac{23}{18480}b^4$ | $-\dfrac{61}{5544}b^5$ |

TABLE 5.3. CONSISTENT MASS MATRIX OF A CURVED STRIP WITH ANY END CONDITIONS

$$[M]^e_{mn} = \int_0^\alpha \varrho t(\theta)\,\Theta_m \Theta_n \, d\theta$$

$$
[M]^e_{mn} =
\begin{bmatrix}
\dfrac{24}{70} b'^2 + \dfrac{26}{35} b' r_i & & & \text{Symmetrical}\\[2ex]
\dfrac{2}{15} b'^3 + \dfrac{8}{105} b'^3 r_i & \dfrac{2}{35} b'^4 + \dfrac{8}{105} b'^3 r_i & & \\[2ex]
\dfrac{9}{35} b'^2 + \dfrac{9}{35} b' r_i & \dfrac{2}{15} b'^3 + \dfrac{13}{105} b'^2 r_i & \dfrac{24}{21} b'^2 + \dfrac{26}{35} b' r_i & \\[2ex]
-\dfrac{4}{35} b'^3 - \dfrac{13}{105} b'^2 r_i & -\dfrac{2}{35} b'^4 - \dfrac{2}{35} b'^3 r_i & -\dfrac{2}{7} b'^3 - \dfrac{22}{105} b'^2 r_i & \dfrac{2}{21} b'^4 + \dfrac{8}{105} b'^3 r_i
\end{bmatrix}
$$

## 5.3.3. CURVED BENDING STRIP (LO2)

This lower order curved strip has been discussed in Section 2.3 and the matrix $[N]_m$ is reproduced below for convenience.

$$[N]_m = \left[ \left(1 - \frac{3}{4} R^2 + \frac{1}{4} R^3\right), \quad b' \left(R - R^2 + \frac{R^3}{4}\right), \quad \left(\frac{3}{4} R^2 - \frac{1}{4} R^3\right), \right.$$

$$\left. b' \left(\frac{R^3}{4} - \frac{R^2}{2}\right) \right] \Theta_m = [C]\Theta_m.$$

Because the polar coordinate system is used here, it is necessary to change (5.7) to a slightly different form,

$$[M]_{mn}^e = \int_{r_1}^{r_2} [C]^T [C] r \, dr \int_0^\alpha \varrho t \Theta_m \Theta_n \, d\theta. \tag{5.9}$$

The algebraic expressions of the matrix coefficients for $[M]_{mn}^e$ are presented in Table 5.3. Again for the special simply supported case with constant $\varrho t$ only the diagonal submatrix $[M]_{mm}^e$ is non-zero and it can be obtained from Table 5.4 by substituting $\varrho t(\alpha/2)$ in place of the integral.

## 5.3.4. SKEW BENDING STRIP (LO2)

In Section 2.4 it was mentioned that the lower order skew plate strip was found to be unsatisfactory for static analysis of skew plates. This, however, has not been the case for vibration analysis, and the lower order skew strip has been successfully applied to the frequency analysis of skew orthotropic plates by Babu and Reddy.[8] Although only the simply supported case has been dealt with in the reference, there should not be any difficulty in extending the analysis to include other types of support conditions.

TABLE 5.4. ELEMENTS OF THE STIFFNESS MATRIX $[S]_{mn}$ FOR THE SKEW STRIP (BABU AND REDDY[8])

| | | | |
|---|---|---|---|
| $\dfrac{6Ga}{b^3}+\dfrac{6}{5}Hk_m^2\dfrac{a}{b}$ $+\dfrac{13}{70}Pk_m^4ab$ $+\dfrac{12}{5}Fk_m^2\dfrac{a}{b}$ | $\dfrac{3Ga}{b^2}+\dfrac{3}{5}Hk_m^2a$ $+\dfrac{11}{420}Pk_m^4ab^2$ $+\dfrac{1}{5}Fk_m^2a$ | $-\dfrac{6Ga}{b^3}-\dfrac{6}{5}Hk_m^2\dfrac{a}{b}$ $+\dfrac{9}{140}Pk_m^4ab$ $-\dfrac{12}{5}Fk_m^2\dfrac{a}{b}$ | $\dfrac{3Ga}{b^2}+\dfrac{1}{10}Hk_m^2a$ $-\dfrac{13}{840}Pk_m^4ab^2$ $+\dfrac{1}{5}Fk_m^2a$ |
| $\dfrac{3Ga}{b^2}+\dfrac{3}{5}Hk_m^2a$ $+\dfrac{11}{420}Pk_m^4ab^2$ $+\dfrac{1}{5}Fk_m^2a$ | $\dfrac{2Ga}{b}+\dfrac{2}{15}Hk_m^2ab$ $+\dfrac{1}{210}Pk_m^4ab^3$ $+\dfrac{4}{15}Fk_m^2ab$ | $-\dfrac{3Ga}{b^2}-\dfrac{1}{10}Hk_m^2a$ $+\dfrac{13}{840}Pk_m^4ab^2$ $-\dfrac{1}{5}Fk_m^2a$ | $\dfrac{Ga}{b}-\dfrac{1}{30}Hk_m^2ab$ $-\dfrac{1}{280}Pk_m^4ab^3$ $-\dfrac{1}{15}Fk_m^2ab$ |
| $-\dfrac{6Ga}{b^3}-\dfrac{6}{5}Hk_m^2\dfrac{a}{b}$ $+\dfrac{9}{140}Pk_m^4ab$ $-\dfrac{12}{5}Fk_m^2\dfrac{a}{b}$ | $-\dfrac{3Ga}{b^2}-\dfrac{1}{10}Hk_m^2a$ $+\dfrac{13}{840}Pk_m^4ab^2$ $-\dfrac{1}{5}Fk_m^2a$ | $\dfrac{6Ga}{b^3}+\dfrac{6}{5}Hk_m^2\dfrac{a}{b}$ $+\dfrac{13}{70}Pk_m^4ab$ $+\dfrac{12}{5}Fk_m^2\dfrac{a}{b}$ | $-\dfrac{3Ga}{b^2}-\dfrac{3}{5}Hk_m^2a$ $+\dfrac{11}{420}Pk_m^4ab^2$ $-\dfrac{1}{5}Fk_m^2a$ |
| $\dfrac{3Ga}{b^2}+\dfrac{1}{10}Hk_m^2a$ | $\dfrac{Ga}{b}-\dfrac{1}{30}Hk_m^2ab$ | $-\dfrac{3Ga}{b^2}-\dfrac{3}{5}Hk_m^2a$ | $\dfrac{2Ga}{b}+\dfrac{2}{15}Hk_m^2ab$ |

$$-\frac{13}{840} Pk_m^4 ab^2 \qquad\qquad -\frac{11}{420} Pk_m^4 ab^2 \qquad\qquad +\frac{1}{210} Pk_m^4 ab^3$$

$$+\frac{1}{5} Fk_m^2 a \qquad\qquad -\frac{1}{5} Fk_m^2 a \qquad\qquad +\frac{4}{15} Fk_m^2 ab$$

$$-\frac{1}{280} Pk_m^4 ab^3$$

$$-\frac{1}{15} Fk_m^2 ab$$

where

$$G = \left[ D_x + \frac{2D_1}{\tan^2\beta} + \frac{D_y}{\tan^4\beta} + \frac{4D_{xy}}{\tan^2\beta} \right] \sin\beta$$

$$H = \left[ \frac{D_1}{\sin^2\beta} + \frac{D_y}{\sin^2\beta\,\tan^2\beta} \right] \sin\beta$$

$$C = \left[ \frac{D_1}{\sin\beta\,\tan\beta} + \frac{2D_{xy}}{\sin\beta\,\tan\beta} + \frac{D_y}{\sin\beta\,\tan^3\beta} \right] \sin\beta$$

$$P = \left[ \frac{D_y}{\sin^4\beta} \right] \sin\beta$$

$$E = \left[ \frac{D_y}{\sin^3\beta\,\tan\beta} \right]$$

$$F = \left[ \frac{D_{xy}}{\sin^2\beta} \right] \sin\beta + \left[ \frac{D_y \cos^2\beta}{\sin^4\beta} \right] \sin\beta$$

and

$$k = m^2\pi^2/a^2.$$

Using skew coordinates $\xi$ and $\eta$, the displacement function for a skew plate strip similar to the one shown in Fig. 2.1f is given as

$$f = w = \sum_{m=1}^{r} [(1 - 3\bar{\xi}^2 + 2\bar{\xi}^3), \quad \xi(1 - 2\bar{\xi} + \bar{\xi}^2), \quad (3\bar{\xi}^2 - 2\bar{\xi}^3), \\ \xi(\bar{\xi}^2 - \bar{\xi})] \sin k_m\eta \begin{Bmatrix} w_{1m} \\ \theta_{1m} \\ w_{2m} \\ \theta_{2m} \end{Bmatrix} \tag{5.10}$$

in which $\bar{\xi} = \xi/b$ and $\theta_{im} = (\partial w/\partial \xi)_{im}$.

Therefore

$$[N]_m = [(1 - 3\bar{\xi}^2 + 2\bar{\xi}^3), \quad \xi(1 - 2\bar{\xi} + \bar{\xi}^2), \quad (3\bar{\xi}^2 - 2\bar{\xi}^3), \quad \xi(\bar{\xi}^2 - \bar{\xi})] \sin k_m\eta. \tag{5.11}$$

The stiffness matrix of this strip can be developed according to the presentation given in Section 2.4 and the mass matrix according to (5.6).

The terms of the series do not couple together in the case of a simply supported strip, and the stiffness and mass matrices for a typical $m$th term are listed in Tables 5.4 and 5.5 respectively.

TABLE 5.5. MASS MATRIX $[M]_{mm}^e$ FOR THE SKEW STRIP (BABU AND REDDY [8])

$$[M]_{mm}^e = \varrho ab \sin \beta \begin{bmatrix} \dfrac{13}{70} & \dfrac{11b}{420} & \dfrac{9}{140} & \dfrac{-13b}{840} \\ \dfrac{11b}{420} & \dfrac{1}{210}b^2 & \dfrac{13}{840}b & \dfrac{-1}{210}b^2 \\ \dfrac{9}{140} & \dfrac{13b}{840} & \dfrac{13}{70} & \dfrac{-11}{420}b \\ \dfrac{-13b}{840} & \dfrac{-1}{210}b^2 & \dfrac{-11}{420}b & \dfrac{1}{210}b^2 \end{bmatrix}$$

## 5.4. CONSISTENT MASS MATRICES FOR PLANE STRESS PLATE STRIPS

A lower order rectangular plate strip was discussed in Section 3.2.1 and the reader might recall that the stiffness formulation was based on the assumption of $Y_m$ for $u$ displacement and $(a/\mu_m)Y'_m$ for $v$ displacement. The same strip will be used here for vibration analysis.

In the previous discussions on flexural vibration problems, the sub-matrix $[M]^e_{mn}$ is always equal to zero for $m \neq n$ provided of course that $\varrho t$ is constant within a strip. However, this is no longer true for plane stress strips because of the existence of $Y'_m$ in the displacement function and hence the non-orthogonal integral $\int_0^a tY'_m Y'_n \, dy$ in the mass matrix. As a result, $[M]^e_{mn}$ will normally remain unequal to zero even if $\varrho t$ is constant.

From (3.7),

$$[N]_m = \begin{bmatrix} (1-\bar{x})Y_m & 0 & \bar{x}Y_m & 0 \\ 0 & (1-\bar{x})\dfrac{a}{\mu_m}Y'_m & 0 & \bar{x}\dfrac{a}{\mu_m}Y'_m \end{bmatrix}.$$

The mass matrix is fairly easy to work out because of the large number of zeros present. The explicit form is given in Table 5.6.

TABLE 5.6. IN-PLANE MASS MATRIX OF A LO2 RECTANGULAR STRIP

$$[M]^e_{mn} = \varrho t$$

| $\dfrac{b}{3}I_1$ | 0 | $\dfrac{b}{6}I_1$ | 0 |
|---|---|---|---|
| 0 | $\left(\dfrac{b}{3C_1C_2}\right)I_2$ | 0 | $\left(\dfrac{b}{6C_1C_2}\right)I_2$ |
| $\dfrac{b}{6}I_1$ | 0 | $\dfrac{b}{3}I_1$ | 0 |
| 0 | $\left(\dfrac{b}{6C_1C_2}\right)I_2$ | 0 | $\left(\dfrac{b}{3C_1C_2}\right)I_2$ |

$$I_1 = \int_0^a Y_m Y_n \, dy, \quad I_2 = \int_0^a Y'_m Y'_n \, dy, \quad C_1 = \frac{\mu_m}{a}, \quad C_2 = \frac{\mu_n}{a}$$

The mass matrix for a curved plane stress plate strip can be derived along the same line and will not be elaborated here. The relevant matrix coefficients are listed in Table 5.7.

TABLE 5.7. IN-PLANE MASS MATRIX OF A LO2 CURVED SLAB STRIP

$$[M]^e_{mn} = \varrho t$$

| $b'(b'+2r_1)I_1/3$ | 0 | $b'(b'+r_1)I_1/3$ | 0 |
|---|---|---|---|
| 0 | $b'(b'+2r_1)I_2/3C_1C_2$ | 0 | $b'(b'+r_1)I_2/3C_1C_2$ |
| $b'(b'+r_1)I_1/3$ | 0 | $b'(3b'+2r_1)I_1/3$ | 0 |
| 0 | $b'(b'+r_1)I_2/3C_1C_2$ | 0 | $b'(3b'+2r_1)I_2/3C_1C_2$ |

$$I_1 = \int_0^\alpha \Theta_m \Theta_n \, d\theta, \quad I_2 = \int_0^\alpha \Theta'_m \Theta'_n \, d\theta, \quad C_1 = \frac{\mu_m}{\alpha}, \quad C_2 = \frac{\mu_n}{\alpha}, \quad r_1 = \text{inner radius.}$$

## 5.5. COMPREHENSIVE MASS MATRIX FOR FLAT SHELL STRIPS

In Chapter 4 it was shown that for a flat shell strip both the bending and in-plane systems of nodal displacements are acting simultaneously and therefore at each nodal line four components of displacement are present. It was further shown that the coefficients of the comprehensive stiffness matrix $[S]_{mn}$ can be made up from appropriate elements of the $[S^p]_{mn}$ and $[S^b]_{mn}$ matrices, where the superscripts $p$ and $b$ refer to plane stress and bending respectively.

For vibration analysis the comprehensive mass matrix is made up in a similar fashion. Let $[M^p]^e_{mn}$ and $[M^b]^e_{mn}$ refer to the in-plane mass matrix and bending mass matrix respectively, and let $[M]^e_{mn}$ represent the comprehensive mass matrix. Then

$$[M]^e_{mn} = \begin{bmatrix} [M_{11}]^e_{mn} & [M_{12}]^e_{mn} \\ [M_{21}]^e_{mn} & [M_{22}]^e_{mn} \end{bmatrix} \qquad (5.12a)$$

and

$$[M_{ij}]^e_{mn} = \begin{bmatrix} [M^p_{ij}]^e_{mn} & [0] \\ [0] & [M^b_{ij}]^e_{mn} \end{bmatrix}. \qquad (5.12b)$$

The transformation procedure presented in Section 4.3 can be used here for the transformation of mass matrices.

## 5.6. VIBRATION OF THIN-WALLED STRUCTURES

The earlier studies on the dynamic behaviour of thin-walled sections were concentrated on stiffened plates, an example being the investigations of Hopmann et al[9, 10] in which stiffened plates were treated as orthotropic plates and the rigidities were derived experimentally. Kirk[11] used isotropic modal shapes in combination with the Raleigh method to determine the frequencies of panels with single or multiple stiffeners.

The finite element vibration analysis of flat thin-walled structures was developed by Anderson.[7] However, due to the large computer capacity and excessive computer time required for performing the analysis, its application to practical problems is severely limited.

A different discrete approach using large elements was proposed by Wittrick *et al.*[12] This approach is similar in concept to the finite strip method except that the stiffness and mass matrices are obtained by solving exactly the plate-bending and plane stress differential equations. Since exact solutions are hard to come by, the method is restricted to isotropic constant thickness plates with an opposite pair of simply supported ends. Also, because of the presence of transcendental functions in the stiffness matrix, a special eigenvalue solution technique has to be used.

The finite strip procedure using flat shell trips (see Section 5.5) is ideally suited for the dynamic analysis of multiple-plate systems because of its ability to deal with a variety of different situations and also because of the small number of unknowns involved. The versatility and accuracy of the method will be amply demonstrated in a number of examples to be presented later.

## 5.7. CYLINDRICAL SHELL STRIP  (Fig. 5.1)

Cylindrical panels are frequently used as structural components for ships, aircraft, and space vehicles which are often subject to dynamic forces, and knowledge of the vibration characteristics of such curved panels is of prime importance.

Several solutions based on the Raleigh–Ritz method[13, 14] are available for panels with all clamped and all simply supported edge conditions. However, the results presented are often valid only for isotropic panels with small curvature and constant thickness. Exact solutions are only known for panels with all sides simply supported.[15]

The above-mentioned problems can be conveniently solved by using circular cylindrical strips with various end conditions along the two straight edges, and the detailed formulation will be presented in this section. For cylindrical panels of other shapes such as elliptical and parabolic cross-sections, the flat shell strip discussed in the previous section can be used so that any arbitrary curve is in fact approximated by a many-sided polygon.

The displacement functions for a cylindrical shell strip are identical to those for a conical frustrum (Fig. 4.3) and, indeed, it is possible to

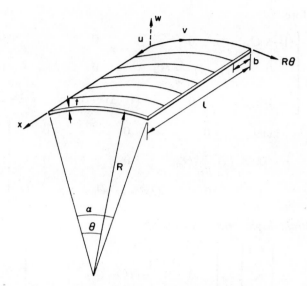

FIG. 5.1. A cylindrical panel and its strip idealization.

obtain the stiffness matrix of a cylindrical shell strip by substituting $\phi = 0°$ into the coefficients of Table 4.2. However, for the sake of clarity both the stiffness matrix and the mass matrix of the cylindrical shell strip will be derived here in detail.

*(i) Displacement functions*

$$
\left.
\begin{aligned}
u &= \sum_{m=1}^{r} [(1-\bar{x}), (\bar{x})]\Theta_m \begin{Bmatrix} u_{1m} \\ u_{2m} \end{Bmatrix}, \\
v &= \sum_{m=1}^{r} [(1-\bar{x}), (\bar{x})] \frac{\alpha}{\mu_m} \Theta'_m \begin{Bmatrix} v_{1m} \\ v_{2m} \end{Bmatrix}, \\
w &= \sum_{m=1}^{r} [(1-3\bar{x}^2+2\bar{x}^3), x(1-2\bar{x}+\bar{x}^2), (3\bar{x}^2-2\bar{x}^3), \\
&\qquad x(\bar{x}^2-\bar{x})] \, \Theta_m \begin{Bmatrix} w_{1m} \\ \psi_{1m} \\ w_{2m} \\ \psi_{2m} \end{Bmatrix}.
\end{aligned}
\right\} \quad (5.13)
$$

Therefore

$$[N]_m = \begin{bmatrix} (1-\bar{x})\Theta_m & 0 & 0 & 0 \\ 0 & (1-\bar{x})\dfrac{\alpha}{\mu_m}\Theta'_m & 0 & 0 \\ 0 & 0 & (1-3\bar{x}^2+3\bar{x}^3)\Theta_m & x(1-2\bar{x}+\bar{x}^2) \\ (\bar{x})\Theta_m & 0 & 0 & 0 \\ 0 & (\bar{x})\dfrac{\alpha}{\mu_m}\Theta'_m & 0 & 0 \\ 0 & 0 & (3\bar{x}^2-2\bar{x}^3)\Theta_m & x(\bar{x}^2-\bar{x})\Theta_m \end{bmatrix} \tag{5.14}$$

*(ii) Strain-displacement relationship*

$$\{\varepsilon\} = \begin{Bmatrix} \varepsilon_x \\ \varepsilon_\theta \\ \gamma_{x\theta} \\ \chi_x \\ \chi_\theta \\ \chi_{x\theta} \end{Bmatrix} = \begin{bmatrix} \partial/\partial x & 0 & 0 \\ 0 & \dfrac{1}{r}\partial/\partial\theta & \dfrac{1}{r} \\ \dfrac{1}{r}\partial/\partial\theta & \partial/\partial x & 0 \\ 0 & 0 & -\partial^2/\partial x^2 \\ 0 & \dfrac{1}{r^2}\partial/\partial\theta & \dfrac{-1}{r^2}\partial^2/\partial\theta^2 \\ 0 & \dfrac{2}{r}\partial/\partial x & \dfrac{-2}{r}\partial^2/\partial x\,\partial\theta \end{bmatrix} \begin{Bmatrix} u \\ v \\ w \end{Bmatrix}$$

$$= \sum_{m=1}^{r} [B]_m \{\delta\}_m. \tag{5.15}$$

The strain matrix $[B]_m$ is given in Table 5.8.

*(iii) Stress–strain relationship*

The stress–strain relationship for a cylindrical panel is the same as the one for a conical shell (Section 4.2.2) and will not be repeated here.

All the ingredients are now complete, and the stiffness matrix $[S]_{mn}$ and mass matrix $[M]^e_{mn}$ can be worked out without difficulty. The stiffness matrix $[S]_{mn}$ will not be listed here since it has exactly the same form as the matrix given in Table 4.2. In fact all the coefficients can be obtained

TABLE 5.8. STRAIN MATRIX OF A CYLINDRICAL SHELL STRIP

$$[B]_m =$$

| $-\frac{1}{b}\Theta_m$ | $0$ | $0$ | $0$ | $\frac{1}{b}\Theta_m$ | $0$ | $0$ | $0$ |
|---|---|---|---|---|---|---|---|
| $0$ | $\frac{1}{C_1 r}\left(1-\frac{x}{b}\right)\Theta''_m$ | $\frac{1}{r}\left(1-\frac{3x^2}{b^2}+\frac{2x^3}{b^3}\right)\Theta_m$ | $\frac{1}{r}\left(x-\frac{2x^2}{b}+\frac{x^3}{b^2}\right)\Theta_m$ | $0$ | $\frac{1}{C_1 r}\left(\frac{x}{b}\right)\Theta''_m$ | $\frac{1}{r}\left(\frac{3x^2}{b^2}-\frac{2x^3}{b^3}\right)\Theta_m$ | $\frac{1}{r}\left(\frac{x^3}{b^2}-\frac{x^2}{b}\right)\Theta_m$ |
| $\frac{1}{r}\left(1-\frac{x}{b}\right)\Theta'_m$ | $\frac{-1}{C_1 b}\Theta'_m$ | $0$ | $0$ | $\frac{1}{r}\left(\frac{x}{b}\right)\Theta'_m$ | $\frac{1}{C_1 b}\Theta'_m$ | $0$ | $0$ |
| $0$ | $0$ | $\left(\frac{6}{b^2}-\frac{12x}{b^3}\right)\Theta_m$ | $\left(\frac{4}{b}-\frac{6x}{b^2}\right)\Theta_m$ | $0$ | $0$ | $\left(\frac{12x}{b^3}-\frac{6}{b^2}\right)\Theta_m$ | $\left(\frac{2}{b}-\frac{6x}{b^2}\right)\Theta_m$ |
| $0$ | $\frac{1}{C_1 r^2}\left(1-\frac{x}{b}\right)\Theta''_m$ | $\frac{-1}{r^2}\left(1-\frac{3x^2}{b^2}+\frac{2x^3}{b^3}\right)\Theta''_m$ | $\frac{-1}{r^2}\left(x-\frac{2x^2}{b}+\frac{x^3}{b}\right)\Theta''_m$ | $0$ | $\frac{1}{C_1 r^2}\left(\frac{x}{b}\right)\Theta''_m$ | $\frac{-1}{r^2}\left(\frac{3x^2}{b^2}-\frac{2x^3}{b^3}\right)\Theta''_m$ | $\frac{-1}{r^2}\left(\frac{x^3}{b^2}-\frac{x^2}{b}\right)\Theta''_m$ |
| $0$ | $\frac{-2}{C_1 r}\left(\frac{1}{b}\right)\Theta'_m$ | $\frac{2}{r}\left(\frac{6x}{b^2}-\frac{6x^2}{b^3}\right)\Theta'_m$ | $\frac{2}{r}\left(\frac{4x}{b}-1-\frac{3x^2}{b^2}\right)\Theta'_m$ | $0$ | $\frac{2}{C_1 r}\left(\frac{1}{b}\right)\Theta'_m$ | $\frac{2}{r}\left(\frac{6x^2}{b^3}-\frac{6x}{b^2}\right)\Theta'_m$ | $\frac{2}{r}\left(\frac{2x}{b}-\frac{3x^2}{b^2}\right)\Theta'_m$ |

$$C_1 - \frac{\mu_m}{\alpha}$$

TABLE 5.9. A CURVED WEB MASS MATRIX

$$[M]^e_{mn} = \varrho t$$

| | | | | | | |
|---|---|---|---|---|---|---|
| $b\left(\dfrac{b}{12}S\phi+\dfrac{r_i}{3}\right)I_1$ | | | | | | |
| | $b\left(\dfrac{b}{12}S\phi+\dfrac{r_i}{3}\right)I_2$ | | | | | |
| | | $b\left(\dfrac{6b}{70}S\phi+\dfrac{13}{35}r_i\right)I_1$ | | | | |
| | | $b^2\left(\dfrac{b}{60}S\phi+\dfrac{11}{210}r_i\right)I_1$ | $b^3\left(\dfrac{b}{280}S\phi+\dfrac{1}{105}r_i\right)I_1$ | | | |
| $b\left(\dfrac{b}{12}S\phi+\dfrac{r_i}{6}\right)I_1$ | | | | $b\left(\dfrac{b}{4}S\phi+\dfrac{r_i}{3}\right)I_1$ | | |
| | $b\left(\dfrac{b}{12}S\phi+\dfrac{r_i}{6}\right)I_2$ | | | $b\left(\dfrac{b}{4}S\phi+\dfrac{r_i}{3}\right)I_2$ | | |
| | | $b\left(\dfrac{9b}{140}S\phi+\dfrac{9}{70}r_i\right)I_1$ | $b^2\left(\dfrac{b}{60}S\phi+\dfrac{13}{420}r_i\right)I_1$ | | $b\left(\dfrac{6b}{21}S\phi+\dfrac{13}{35}r_i\right)I_1$ | |
| | | $-b^2\left(\dfrac{b}{70}S\phi+\dfrac{13}{420}r_i\right)I_1$ | $-b^3\left(\dfrac{b}{280}S\phi+\dfrac{1}{140}r_i\right)I_1$ | | $-b^2\left(\dfrac{b}{28}S\phi+\dfrac{11}{210}r_i\right)I_1$ | $b^3\left(\dfrac{b}{168}S\phi+\dfrac{1}{105}r_i\right)I_1$ |

$$S\phi = \sin\phi; \quad I_1 = \int_0^\alpha \Theta_m \Theta_n \, d\theta; \quad I_2 = \frac{\alpha^2}{\mu_m \mu_n} \int_0^\alpha \Theta'_m \Theta'_n \, d\theta.$$

from Table 4.2 by using the appropriate $B_{ij}$ from Table 5.8 and by putting in the correct integral

$$[S]_{mn} = R \int_0^\alpha \int_0^b [\bar{S}]_{mn} \, dx \, d\theta. \tag{5.16}$$

The mass matrix $[M]_{mn}^e$ is much simpler in form and is listed in Table 5.10.

TABLE 5.10. MASS MATRIX OF A CYLINDRICAL SHELL STRIP

$$R\varrho t \begin{bmatrix}
\dfrac{b}{3}I_1 & 0 & 0 & 0 & \dfrac{b}{6}I_1 & 0 & 0 & 0 \\[2ex]
0 & \left(\dfrac{b}{3C_1C_2}\right)I_2 & 0 & 0 & 0 & \left(\dfrac{b}{6C_1C_2}\right)I_2 & 0 & 0 \\[2ex]
0 & 0 & \dfrac{78b}{210}I_1 & \dfrac{11b^2}{210}I_1 & 0 & 0 & \dfrac{9b}{70}I_1 & \dfrac{-13b^2}{420}I_1 \\[2ex]
0 & 0 & \dfrac{11b^2}{210}I_1 & \dfrac{b^3}{105}I_1 & 0 & 0 & \dfrac{13b^2}{420}I_1 & \dfrac{-3b^3}{420}I_1 \\[2ex]
\dfrac{b}{6}I_1 & 0 & 0 & 0 & \dfrac{b}{3}I_1 & 0 & 0 & 0 \\[2ex]
0 & \left(\dfrac{b}{6C_1C_2}\right)I_2 & 0 & 0 & 0 & \left(\dfrac{b}{3C_1C_2}\right)I_2 & 0 & 0 \\[2ex]
0 & 0 & \dfrac{9b}{70}I_1 & \dfrac{13b^2}{420}I_1 & 0 & 0 & \dfrac{13b}{35}I_1 & \dfrac{-11b^2}{420}I_1 \\[2ex]
0 & 0 & \dfrac{-13b^2}{420}I_1 & \dfrac{-3b^3}{420}I_1 & 0 & 0 & \dfrac{-11b^2}{420}I_1 & \dfrac{b^3}{105}I_1
\end{bmatrix}$$

$$I_1 = \int_0^\alpha \Theta_m \Theta_n \, d\theta, \quad I_2 = \int_0^\alpha \Theta'_m \Theta'_n \, d\theta, \quad C_1 = \frac{\mu_m}{\alpha}, \quad C_2 = \frac{\mu_n}{\alpha}.$$

As a matter of interest, the mass matrix for a conical shell strip is given in Table 5.9. The reader can verify that by specifying $\phi = 0°$ the coefficients in the two tables (5.9 and 5.10) will become identical.

## 5.8. NUMERICAL EXAMPLES

(1) The natural frequencies of several square isotropic plates with different boundary conditions were obtained by using (a) eight LO2 rectangular bending strips for the whole plate and four terms of the series, and (b) two HO2 strips and two terms.[17] The results of the two analyses were compared in Table 5.11 with those due to Warburton[18] and very little

TABLE 5.11. NATURAL FREQUENCIES OF SLABS WITH VARIOUS BOUNDARY CONDITIONS

– – – Simply supported    ▬▬ Clamped    ——— Free

| Frequencies | HO2 | LO2 | Ref. (18) | HO2 | LO2 | Ref (18) | HO2 | LO2 | Ref. (18) | HO2 | LO2 | Ref. (18) |
|---|---|---|---|---|---|---|---|---|---|---|---|---|
| $\omega_1$ | 19·74 | 19·74 | 19·74 | 23·66 | 23·62 | 23·77 | 36·05 | 36·01 | 36·00 | 22·33 | 22·29 | 22·38 |
| $\omega_2$ | 49·35 | 49·32 | 49·35 | 51·79 | 51·62 | 52·00 | 73·44 | 73·48 | 73·41 | 27·19 | 27·08 | 27·33 |
| $\omega_3$ | 49·36 | 49·34 | 49·35 | 58·66 | 58·65 | 58·65 | 73·75 | 73·96 | 73·41 | 45·19 | 44·76 | 45·49 |
| $\omega_4$ | 78·97 | 78·91 | 78·95 | 86·41 | 86·16 | 86·26 | 108·53 | 108·91 | 108·24 | 61·55 | 61·53 | 61·68 |
| $\omega_5$ | 98·94 | 98·64 | 98·69 | 100·89 | 100·35 | 10084 | 13353 | 132·09 | 131·90 | 68·33 | 68·29 | 6873 |

difference can be found between the two sets of results. Note that four terms of the series were used in (a) because eight frequencies of high accuracy were originally required and only four frequencies were quoted here.

(2) The LO2 rectangular bending strip is now applied to a variable thickness plate problem, and the accuracy attainable for this type of structure is tested against the solution of Appl and Byers[19] who only calculated the fundamental frequency of simply supported plates with linearly varying thickness in one direction. The finite strip method, of course, can give as many frequencies as required for a variety of boundary conditions. A square plate with a thickness variation of 1.0–1.8 (Fig. 5.2) is analysed by using two different types of strips. The first type is a standard constant thickness strip and the plate is approximated by a series of strips each with different thicknesses, while the second type is a strip with linear variable thickness in the longitudinal direction and constant thickness in the transverse direction, so that the plate is now made up of identical strips. It can be seen from Table 5.12 that both sets of results compared favourably with each other. The fundamental frequency from both analyses also agree quite well with the result given in ref. 19.

Note that in the second case, due to the fact that the bending rigidities and $\varrho t$ of the strip are now functions of $y$, the stiffness matrix $[S]_{mn}$ and

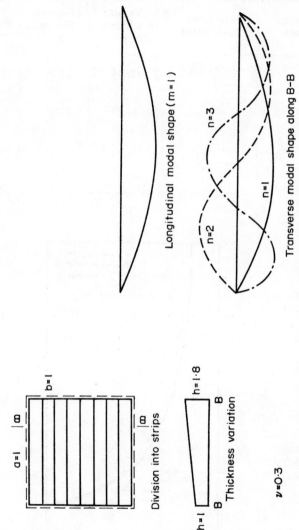

Fig. 5.2.  Simply supported variable thickness plate with some modal shapes.

TABLE 5.12. FREQUENCIES OF VARIABLE THICKNESS PLATE-
MULTIPLIER $(1/a^2)$

| Longitudinal modal shape | Transverse modal shape | Variable thickness strip | Constant thickness strip | Ref. 19 |
|---|---|---|---|---|
| $(m=1)$ | $(n=1)$ | 27.45 | 27.35 | 27.35 |
| $(m=1)$ | $(n=2)$ | 68.20 | 67.99 | |
| $(m=1)$ | $(n=3)$ | 136.39 | 135.61 | |

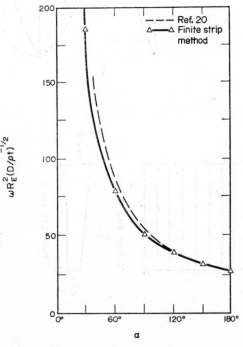

FIG. 5.3. Variation of lowest frequency with sectorial angle.

the mass matrix $[M]_{mn}^e$ for $m \neq n$ will no longer be equal to zero although the strip is simply supported at the two ends.

(3) The LO2 curved bending strip is used to analyse clamped sector plates and the results are compared with those computed by Ben-Amoz[20] (Fig. 5.3). The agreement is very good on the whole, and the frequencies obtained by the two methods are in fact identical for the

TABLE 5.13. EIGENVALUES ($\lambda' = \omega\sqrt{p/D}$) FOR SKEW ISOTROPIC PLATES

| Span width | Skew angle ($\beta'$) | Finite strip $\nu = 0.3$ | | | Periasamy[21] $\nu = 0.15$ | | |
|---|---|---|---|---|---|---|---|
| | | Mode 1 | Mode 2 | Mode 3 | Mode 1 | Mode 2 | Mode 3 |
| 0.5 | 30° | 12.78 | 16.50 | 26.64 | 12.80 | 15.22 | 23.52 |
| | 45° | 18.78 | 25.94 | 44.01 | 19.46 | 22.20 | 34.08 |
| 1.0 | 30° | 12.47 | 24.37 | 55.05 | 12.58 | 20.25 | 52.47 |
| | 45° | 17.96 | 40.30 | 90.71 | 18.82 | 27.24 | 81.76 |

Babu and Reddy[8]

range of subtended angles from 120 to 180. Eight strips and four terms of the series are used in this analysis. It was mentioned in Chapter 2 that no numerical instability was ever experienced in the bending analysis of a clamped sector plate although the input data for the inner radius of the first strip is equal to zero. The same statement can be made for the vibration analysis of sector plates.

(4) Isotropic skew bridges with different skew angles and aspect ratios were computed by Babu and Reddy,[8] using the LO2 skew strip discussed in Section 5.3.4. The results compared quite well with Periasamy's values (Table 5.13).

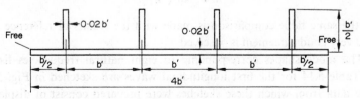

FIG. 5.4. A simply supported four–stiffener panel ($l = 6b'$).

TABLE 5.14. NATURAL FREQUENCIES OF A FOUR STIFFENER PANEL IN THE
RANGE $0 < \bar{n} < 0.081$ (12 STRIPS)

| Mode number | $\bar{n} = \dfrac{\omega l}{\sqrt{E/\varrho}}$ | | Wave number (m) | Type of symmetry |
|:---:|:---:|:---:|:---:|:---:|
| | FS | Reference (12) | | |
| 1 | 0.0287 | 0.0286 | 1 | A |
| 2 | 0.0292 | 0.0291 | 1 | S |
| 3 | 0.0365 | 0.0359 | 1 | S |
| 4 | 0.0366 | 0.0362 | 1 | A |
| 5 | 0.0394 | 0.0391 | 1 | S |
| 6 | 0.0396 | 0.0395 | 2 | S |
| 7 | 0.0411 | 0.0410 | 2 | A |
| 8 | 0.0504 | 0.0504 | 3 | S |
| 9 | 0.0521 | 0.0519 | 3 | A |
| 10 | 0.0557 | 0.0555 | 2 | S |
| 11 | 0.0639 | 0.0636 | 1 | A |
| 12 | 0.0643 | 0.0641 | 4 | S |
| 13 | 0.0647 | 0.0642 | 1 | S |
| 14 | 0.0648 | 0.0645 | 2 | A |
| 15 | 0.0653 | 0.0647 | 1 | A |
| 16 | 0.0653 | 0.0649 | 3 | S |
| 17 | 0.0656 | 0.0654 | 4 | A |
| 18 | 0.0736 | 0.0731 | 3 | A |
| 19 | 0.0773 | 0.0768 | 4 | S |
| 20 | 0.0808 | 0.0804 | 5 | S |

(5) A four-stiffener panel (Fig. 5.4) with two ends simply supported and the other edges free was analysed by using the flat shell strip described in Section 5.5. The Poisson ratio $v$ is assumed to be 0.3 throughout. The lowest twenty natural frequencies of the panel are presented in Table 5.14, in terms of a dimensionless parameter $\bar{n}$ defined by

$$\bar{n} = \omega a/\sqrt{E/\varrho}.$$

In the same table comparison is made with the results of reference 12, and very good agreement is observed.

The mode shapes corresponding to eight natural frequencies listed in Table 5.14 for the first longitudinal waves are sketched in Fig. 5.5. The data from which these sketches were prepared consist of displacements and rotations at the top and bottom of all four stiffeners. It should

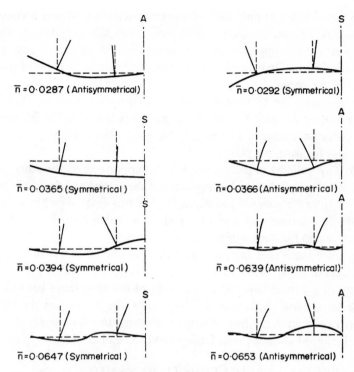

FIG. 5.5. Modal shapes for the first eight natural frequencies of a four-stiffener panel at a wave number $m = 1$.

TABLE 5.15. NATURAL FREQUENCIES (RADIAN/SEC) OF CYLINDRICAL PANELS

| Panel shape and boundary | | | | |
|---|---|---|---|---|
| Frequencies | Curved strip | Flat strip | Curved strip | Flat strip |
| $\omega_1$ | 0.277 (0.282)[15] | 0.285 | 0.101 (0.108)[13] | 0.100 |
| $\omega_2$ | 0.277 | 0.305 | 0.127 | 0.119 |
| $\omega_3$ | 0.466 | 0.512 | 0.139 | 0.142 |
| $\omega_4$ | 0.493 | 0.530 | 0.166 | 0.177 |
| $\omega_5$ | 0.542 | 0.573 | 0.196 | 0.206 |

be pointed out that only half of the cross-section is shown because of symmetry or anti-symmetry conditions. Since the structure is simply supported at two opposite ends, all the longitudinal modes are uncoupled and only a small matrix of fifty-two equations has to be solved for each term of the series.

It is interesting to note that all the frequencies given by the finite strip method in Table 5.14 constitute an upper bound to the frequencies computed in reference 12 through the use of exact stiffness and mass matrices for isotropic plates.

(6) Two cylindrical panels, one simply supported and the other clamped, were analysed using both cylindrical shell strips and flat shell strips. The circular frequencies computed for the two cases together with the results of references 13 and 15 are given in Table 5.15. For each case six strips and four terms of the series are used.

It can be seen that the agreement between the three sets of results is quite good. However, from experience it has been found that the idealization of a curved panel as an assembly of flat shell strips is more suitable for shallow shells than for deep shells, due to the fact the for the latter case a great number of strips is required to approximate the geometric shape before accurate frequencies can be obtained.

## 5.9. SPECIAL APPLICATION TO DYNAMIC ANALYSIS OF BRIDGES

In any bridge, apart from the consideration of the deflections and stresses caused by static loadings, very often it is necessary to consider the dynamic response due to external excitations such as earthquake or moving vehicles. The dynamic stresses and deflections induced by such external excitations can be computed accurately once the natural frequencies and the mode shapes of the structure are known. In other cases, it is necessary to check the natural frequencies simply to avoid the possibility of resonance.

The finite strip method is a powerful tool not only for the static analysis, but also for the dynamic analysis, of various types of bridges. Cheung et al.[22] used a LO2 rectangular bending strip for the analysis of single and continuous span orthotropic slab bridges, while Babu and Reddy[8] studied the free vibration of skew bridges for a variety of skew angles and aspect ratios. The free vibration of right and curved box girder bridges was presented in a paper by Cheung and Cheung.[23] Dynamic deflections and stresses in slab bridges due to moving loads were computed by Smith.[24]

Two examples on the vibration of bridges will be presented here. The first example is on the analysis of a continuous, variable thickness deck (Fig. 5.6a). Such structures are frequently met with in practice, and they can be conveniently idealized as an assembly of transverse strips such that each strip should have its own individual thickness (Fig. 5.6b).

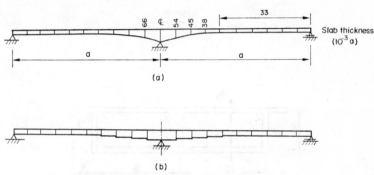

FIG. 5.6. (a) Longitudinal section. (b) Idealized section. Circular frequencies 3.40, 6.76, 7.58, 9.30, 14.46.

$$\left[ \text{Unit:} \ \frac{1}{100^2} \sqrt{\frac{H}{\varrho}}, \ H = \frac{E}{12(1-\nu^2)} \right].$$

Note that the strips used for this example are completely free at the two ends $[Y_m$ of (1.6d)], and that by increasing the number of strips it is possible to analyse bridges with any number of spans. At each support the deflection parameters for all the terms of the series are simply suppressed.

The second example presented here is a twin-cell curved box girder bridge shown in Fig. 5.7A. It is worth while mentioning that the midpoint nodal lines of the three webs (Fig. 5.7B) can all be removed without prejudicing the accuracy of the solution, and that only fifty-six equations per harmonic would be quite adequate for solving such a complicated structure. The computation time for thirty-six frequencies (nine for each harmonic) is only about 4 min on the medium speed IBM 360–50 computer.

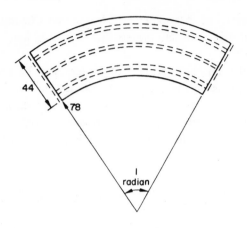

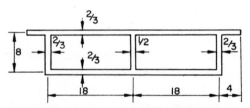

FIG. 5.7A. Plan and section of the curved box girder bridge ($E = 1$, $\nu = 0.16$).

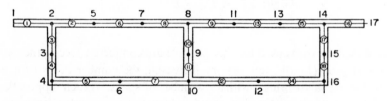

FIG. 5.7B. Numbering of nodal lines and strips for bridge.

Several frequencies and their corresponding mode shapes are shown in Fig. 5.7c.

The bridge is then subject to a concentrated mass attached to the top of the mid-section of the outer web. This concentrated mass, which corresponds to the presence of a heavy stationary vehicle, is assumed to be equal to one-eighth of the total mass of the structure. From Table 5.16 it is possible to conclude that the additional mass will, in general, lower

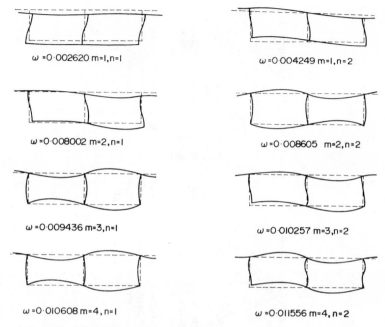

$\omega =0.002620$ m=1,n=1                    $\omega =0.004249$ m=1,n=2

$\omega =0.008002$ m=2,n=1                    $\omega =0.008605$ m=2,n=2

$\omega =0.009436$ m=3,n=1                    $\omega =0.010257$ m=3,n=2

$\omega =0.010608$ m=4,n=1                    $\omega =0.011556$ m=4,n=2

FIG. 5.7c. Modal shapes of a curved box girder bridge ($E = 1$, $\nu = 0.16$, $\varrho = 1$).

the natural frequencies of the bridge. However, if the concentrated mass is placed on or near a nodal line (in vibration terminology this means a line along which the masses are stationary), there will be little or no effect on the values of the frequencies. For example, no change can be observed for frequencies which correspond to the anti-symmetric modes.

## 5.10. STABILITY ANALYSIS OF STIFFENED PLATE STRUCTURES

Comparatively little work has been done on the application of the finite strip to stability problems, although vibration and stability share many similar features and both require the determination of eigenvalues and eigenvectors. Yoshida[25] used rectangular shell strips in combination with eccentric beams to predict the critical loads of stiffened plates, while a study on the influence of orthotropy on the stability of multiple-

TABLE 5.16. NATURAL FREQUENCIES (RADIAN/SEC) OF A CURVED BOX GIRDER BRIDGE WITH CONCENTRATED MASS AT MID-SECTION OF OUTER WEB ($E = 1$, $\nu = 0.16$, $\varrho = 1$)

| Circular frequencies | $\omega_1$ | | $\omega_2$ | | $\omega_3$ | | $\omega_4$ | | $\omega_5$ | |
|---|---|---|---|---|---|---|---|---|---|---|
| Longitudinal mode | With concentrated mass | Without concentrated mass | With concentrated mass | Without concentrated mass | With concentrated mass | Without concentrated mass | With concentrated mass | Without concentrated mass | With concentrated mass | Without concentrated mass |
| $m = 1$ | 0.002283 | 0.002620 | 0.003742 | 0.004249 | 0.007965 | 0.008066 | 0.008073 | 0.008501 | 0.008503 | 0.008553 |
| $m = 2$ | 0.008002 | 0.008002 | 0.008605 | 0.008605 | 0.009107 | 0.009107 | 0.011427 | 0.011427 | 0.013644 | 0.013644 |
| $m = 3$ | 0.009434 | 0.009436 | 0.010016 | 0.010257 | 0.011700 | 0.011821 | 0.012240 | 0.012246 | 0.015050 | 0.020046 |
| $m = 4$ | 0.010608 | 0.010608 | 0.011556 | 0.011556 | 0.013459 | 0.013459 | 0.013523 | 0.013523 | 0.025243 | 0.025243 |

plate structures were reported by Turvey[26] and by Turvey and Wittrick.[27] The discussions presented here are based on the work of Yoshida.

Consider the simply supported LO2 flat shell strip shown in Fig. 4.1. The strip is subject to an initial stress $\sigma$ which varies linearly from side 1 to side 2, but is constant along the longitudinal axis. The potential energy due to the in-plane forces is given by the expression

$$\frac{t}{2} \int_0^a \int_0^b \{\sigma_1 - (\sigma_1 - \sigma_2)\bar{x}\} \left\{ \left(\frac{\partial u}{\partial y}\right)^2 + \left(\frac{\partial v}{\partial y}\right)^2 + \left(\frac{\partial w}{\partial y}\right)^2 \right\} dx \, dy. \quad (5.17)$$

The quadratic terms inside (5.17) can be written as

$$\begin{bmatrix} \dfrac{\partial u}{\partial y} & \dfrac{\partial v}{\partial y} & \dfrac{\partial w}{\partial y} \end{bmatrix} \begin{Bmatrix} \dfrac{\partial u}{\partial y} \\[2mm] \dfrac{\partial v}{\partial y} \\[2mm] \dfrac{\partial w}{\partial y} \end{Bmatrix}. \quad (5.18)$$

Noting that the slopes are related to the nodal displacement parameters of the strip, we can write

$$\begin{Bmatrix} \dfrac{\partial u}{\partial y} \\[2mm] \dfrac{\partial v}{\partial y} \\[2mm] \dfrac{\partial w}{\partial y} \end{Bmatrix} = [G]\{\delta\} \quad (5.19)$$

and the additional potential energy for the whole strip thus becomes simply

$$\frac{1}{2}\{\delta\}^T [S_G]\{\delta\} \quad (5.20a)$$

with

$$[S_G] = t \int_0^a \int_0^b \{\sigma_1 - (\sigma_1 - \sigma_2)\bar{x}\} [G]^T [G] \, dx \, dy \quad (5.20b)$$

It is obvious that if the total potential energy of the strip, i.e. the sum of strain energy due to bending, potential energy due to nodal line forces,

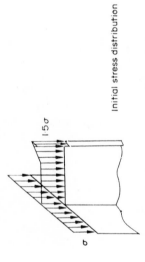

Initial stress distribution

TABLE 5.17. COMPARISON OF FINITE STRIP METHOD WITH FINITE ELEMENT METHOD (YOSHIDA[25])

| Method | Shape of element | Number of elements | Number of free variables | Aspect ratio | Bending rigidity of stiffener | Area of stiffener | Buckling coefficient | Computing time (sec) |
|---|---|---|---|---|---|---|---|---|
| FSM | Strip element + beam element | 12 | 24 | 2.5 | 14.25 | 0.0833 | 11.14 | 1.4 |
| FEM | Triangular element + beam element | 320 | 451 | 2.5 | 14.25 | 0.0838 | 11.08 | 110 |

and the additional potential energy due to the initial stress, is now minimized with respect to the nodal displacement parameters, the following relationship would be obtained:

$$[S]\{\delta\}+[S_G]\{\delta\} = \{P\} \tag{5.21}$$

in which $[S_G]$ is referred to as the geometric stiffness matrix or initial stress matrix and takes up the same sign as the stresses.

Upon assembly of the contributions from all the strips an overall set of equilibrium equations is established,

$$[K]\{\delta\}+[K_G]\{\delta\} = \{P\}. \tag{5.22}$$

For linear stability, the nodal forces are zero and it is therefore possible to arrive at eigenvalue equations similar to the ones given in (5.3),

$$([K]+\lambda[K_G])\{\delta\} = \{0\} \tag{5.23}$$

with $\lambda$ being scaling factor related to the critical load.

The geometric stiffness matrix of an eccentric stiffener has also been worked out by Yoshida in reference 25.

As an illustrative example the simple case of a plate with one concentric stiffener was analysed by both finite strip method and finite element method, using the non-conforming triangular bending element.[28] The results are presented in Table 5.17, and it can be concluded that while the accuracy attainable is about the same for both methods the disparity in computer time is so striking that the finite strip method is undoubtedly much more suitable for this class of problems. Indeed, the problem of a plate with several stiffeners becomes intractable for the finite element method because of the very large computer core requirements.

## REFERENCES

1. J. S. ARCHER, Consistent mass matrix for distributed systems, *Am. Soc. Civ. Engrs* **89**, ST4, 161 (1963).
2. F. A. LECKIE and G. M. LINDBERG, The effect of lumped parameters on beam frequencies, *Aeronaut. Quarterly* **14**, 234 (1963).
3. O. C. ZIENKIEWICZ and Y. K. CHEUNG, The finite element method for analysis of elastic isotropic and orthotropic slabs, *Proc. Inst. Civ. Engrs* **28**, 471 (1964).
4. R. E. D. BISHOP, G. M. L. GLADWELL, and S. MICHAELSON, *The Matrix Analysis of Vibration*, Cambridge University Press, 1965.

5.  B. M. IRONS, Eigenvalue economisers in vibration problems, *J. Aeronaut. Soc.* **67**, 526 (1963).
6.  K. J. BATHE, *Solution Methods for Large Generalized Eigenvalue Problems in Structural Engineering*, SESM Report 71–20, Department of Civil Engineering, University of California, Berkeley, 1971.
7.  R. G. ANDERSON, A finite element eigenvalue solution system, PhD thesis, University of Wales, Swansea, 1968.
8.  P. V. T. BABU and D. V. REDDY, Frequency analysis of skew orthotropic plates by the finite strip method, *J. Sound Vib.* **18**, 465–74 (1971).
9.  W. H. HOPPMANN, N. J. HUFFINGTON, and L. S. MAGNESS, A study of orthogonally stiffened plates., *J. appl. Mech., Trans. ASME* **23**, 343–50 (1956).
10. W. H. HOPPMANN and L. S. MAGNESS, Nodal patterns of the free flexural vibration of stiffened plates, *J. appl. Mech., Trans. ASME* **24**, 526–30 (1957).
11. C. L. KIRK, Vibration characteristics of stiffened plates, *J. Mech. Engng Sci.* **2**, 242–53 (1960).
12. W. H. WITTRICK and F. W. WILLIAMS, Natural vibrations of thin, prismatic, flat-wall structures, *IUTAM Symposium on High Speed Computing of Elastic Structures, Liége, August 1970*.
13. N. R. MADDOX, H. E. PLUMBLEE, and W. W. KING, Frequency analysis of a cylindrically curved panel with clamped and elastic boundaries, *J. Sound Vib.* **12**, 225–49 (1970).
14. J. L. SEWALL, *Vibration Analysis of Cylindrically Curved Panels with Simply Supported or Clamped Edges and Comparison with Some Experiments*, NASA TN D–3791.
15. R. CHEN, *Vibration Analysis of Cylindrical Panels and Rectangular Plates Carrying a Concentrated Mass*, Douglas Aircraft Co. Inc., California, Report, 1965.
16. Y. K. CHEUNG and M. S. CHEUNG, Flexural vibration of rectangular and other polygonal plates, *Am. Soc. Civ. Engrs* **97**, EM2, 391–411 (April 1971).
17. M. S. CHEUNG and Y. K. CHEUNG, Static and dynamic behaviour of rectangular plates using higher order finite strips, *Building Sci.* **7** (3) 415–19 (July 1972).
18. G. B. WARBURTON, The vibration of rectangular plates, *Proc. Instn Mech. Engrs* **168** (12) 371–84 (1954).
19. F. C. APPL and N. R. BYERS, Fundamental frequency of simply supported rectangular plates with linearly varying thickness, *J. Appl. Mech., Trans. ASME* **32**, 163–8 (1965).
20. M. BEN-AMOZ, Note on deflections and flexural vibrations of clamped sectorial plates, *J. App. Mech., Trans. ASME* **26**, 136 (1959).
21. K. PERIASAMY, Vibration of skew plates, ME thesis, Indian Institute of Science, Bangalose, 1969.
22. Y. K. CHEUNG, M. S. CHEUNG, and D. V. REDDY, Frequency analysis of certain single and continuous span bridges, *Development in Bridge Design Construction* (ed. Rockey *et al.*), Crosby Lockwood, 1972.
23. Y. K. CHEUNG and M. S. CHEUNG, Free vibration of curved and straight beam-slab or box-girder bridges, *Publications, International Association for Bridges and Structural Engineering*, 32-II, 41–52, 1972.
24. J. W. SMITH, Finite strip analysis of the dynamic response of beam and slab highway bridges, *Earthquake Engineering and Structural Dynamics*, **4**, 357–70 (1973).
25. KOICHIRO YOSHIDA, Buckling analysis of plate structures by strip elements, *Proc. Jap. Soc. of Naval Architects* **130** (1971).

26. G. J. TURVEY, A contribution to the elastic stability of thin walled structures fabricated from isotropic and orthotropic materials, PhD thesis, Department of Civil Engineering, University of Birmingham, 1971.
27. G. J. TURVEY and W. H. WITTRICK, The influence of orthotropy on the stability of some multi-plate structures in compression, *Aeronaut. Quarterly* **24**, 1–8 (February 1973).
28. G. P. BAZELEY, Y. K. CHEUNG, B. M. IRONS, and O. C. ZIENKIEWICZ, Triangular elements in bending—conforming and non-conforming solutions, *Proceedings of the Conference on Matrix Methods in Structural Mechanics, Air Force Institute of Technology, Wright Patterson Air Force Base, Ohio, October, 1965.*

# CHAPTER 6

# *Further developments in finite strip analysis*

## 6.1. INTRODUCTION

In the previous chapters the finite strip method has been applied to bending, vibration, and stability analysis of ordinary thin plates and shells with standard support conditions. Here the method will be further extended to cover the analysis of plates and shells with special support conditions and also of conventional sandwich plates and multi-layer sandwich plates.

## 6.2. FREE VIBRATION OF PARTIALLY CLAMPED PLATES USING A "MIXED" STRIP[1]

A partially clamped rectangular plate is shown in Fig. 6.1 and it can be seen that both simply supported and clamped conditions exist along the same boundary. Theoretically speaking both the finite element method and finite difference method can be used to deal with this type of mixed boundary problems, although in practice they are very often uneconomical in terms of computational cost.

The finite strip method provides a convenient tool for the accurate analysis of rectangular or sector plates with an opposite pair of partially clamped sides. The solution procedure is started in the same way as in a standard finite strip analysis in which, first of all, a plate is divided into strips with preset end conditions which may differ from strip to strip (Fig. 6.1). However, at the point where an abrupt change occurs it is not possible to satisfy the boundary condition exactly, and a so-called "mixed" strip has to be introduced. This special strip is formulated by using two basic functions to satisfy the two different boundary conditions at the two nodal lines of the strip, and within these two sides the boundary

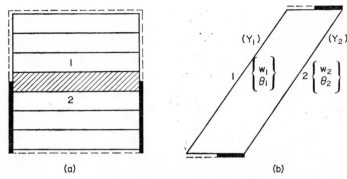

FIG. 6.1. (a) A partially clamped plate and its strip idealization. (b) A mixed strip.

conditions undergo a gradual change from one type to the other. In order to approximate the sudden change closely, such "mixed" strips are usually made as narrow as possible.

Consider a typical "mixed" strip shown in Fig. 6.1. A suitable displacement function for the rectangular strip is

$$w = \sum_{m=1}^{r} \left[ [C_1] \, (Y_1)_m \, [C_2] \, (Y_2)_m \right] \{\delta\}_m \tag{6.1a}$$

in which $(Y_i)_m$ represents the $Y_m$ at the $i$th edge and

$$\{\delta\}_m = [w_{1m}\theta_{1m}w_{2m}\theta_{2m}]^T.$$

The shape functions $[C_1]$ and $[C_2]$ are given by (1.9b). By definition (see Section 1.3.2) a shape function takes up a value of unity at its own nodal line and a value of zero at the other nodal lines. Therefore at side $1(x = 0)$, we have

$$w = \sum_{m=1}^{r} \left[ [1 \ 0] \, (Y_1)_m \, [0 \ 0] \, (Y_2)_m \right] \{\delta\}_m \tag{6.1b}$$

and at side $2(x = b)$, we have

$$w = \sum_{m=1}^{r} \left[ [0 \ 0] \, (Y_1)_m \, [1 \ 0] \, (Y_2)_m \right] \{\delta\}_m. \tag{6.1c}$$

Equation (6.1b) simply states that at side 1, only $(Y_1)_m$ which satisfies the end conditions there is present. A similar statement applies to (6.1c).

TABLE 6.1. STIFFNESS MATRIX OF A RECTANGULAR MIXED STRIP

$$[s]_{mn} =$$

| | | | |
|---|---|---|---|
| $+\dfrac{12D_x}{b^3}I_1 - \dfrac{6D_1}{5b}I_3$<br>$-\dfrac{6D_1}{5b}I_2 + \dfrac{13bD_y}{35}I_4$<br>$+\dfrac{24D_{xy}}{5b}I_5$ | $+\dfrac{6D_x}{b^2}I_1 - \dfrac{11D_1}{10}I_3$<br>$-\dfrac{D_1}{10}I_2 + \dfrac{11b^2D_y}{210}I_4$<br>$+\dfrac{2D_{xy}}{5}I_5$ | $-\dfrac{12D_x}{b^3}I_6 + \dfrac{6D_1}{5b}I_{10}$<br>$+\dfrac{6D_1}{5b}I_7 + \dfrac{9bD_y}{70}I_8$<br>$-\dfrac{24D_{xy}}{5b}I_9$ | $+\dfrac{6D_x}{b^2}I_6 - \dfrac{D_1}{10}I_{10}$<br>$-\dfrac{D_1}{10}I_7 - \dfrac{13b^2D_y}{420}I_8$<br>$+\dfrac{2D_{xy}}{5}I_9$ |
| $+\dfrac{6D_x}{b^2}I_1 - \dfrac{11D_1}{10}I_2$<br>$\dfrac{D_1}{10}I_3 + \dfrac{11b^2D_y}{210}I_4$<br>$+\dfrac{2D_{xy}}{5}I_5$ | $+\dfrac{4D_x}{b}I_1 - \dfrac{2bD_1}{15}I_3$<br>$-\dfrac{2bD_1}{15}I_2 + \dfrac{2b^3D_y}{210}I_4$<br>$+\dfrac{8bD_{xy}}{15}I_5$ | $-\dfrac{6D_x}{b^2}I_6 + \dfrac{D_1}{10}I_{10}$<br>$+\dfrac{D_1}{10}I_7 + \dfrac{26b^2D_y}{840}I_8$<br>$+\dfrac{2D_{xy}}{5}I_9$ | $+\dfrac{2D_x}{b}I_6 + \dfrac{bD_1}{30}I_{10}$<br>$+\dfrac{bD_1}{30}I_7 - \dfrac{2b^3D_y}{280}I_8$<br>$-\dfrac{2bD_{xy}}{15}I_9$ |
| $-\dfrac{12D_x}{b^3}I_6 + \dfrac{6D_1}{5b}I_8$<br>$+\dfrac{6D_1}{5b}I_{10} + \dfrac{9bD_y}{70}I_8$<br>$-\dfrac{24D_{xy}}{5b}I_9$ | $-\dfrac{6D_x}{b^2}I_6 + \dfrac{D_1}{10}I_7$<br>$+\dfrac{D_1}{10}I_{10} + \dfrac{26b^2D_y}{840}I_8$<br>$-\dfrac{2D_{xy}}{5}I_9$ | $+\dfrac{12D_x}{b^3}I_{11} - \dfrac{6D_1}{5b}I_{13}$<br>$+\dfrac{6D_1}{5b}I_{12} + \dfrac{13bD_y}{35}I_{14}$<br>$+\dfrac{24D_{xy}}{5b}I_{15}$ | $-\dfrac{6D_x}{b^2}I_{11} + \dfrac{22D_1}{20}I_{13}$<br>$+\dfrac{2D_1}{20}I_{12} - \dfrac{11b^2D_y}{210}I_{14}$<br>$-\dfrac{2D_{xy}}{5}I_{15}$ |

$$+\frac{6D_x}{b^2}I_6 - \frac{D_1}{10}I_7 \qquad +\frac{2D_x}{b}I_6 + \frac{bD_1}{30}I_7 \qquad -\frac{6D_x}{b^2}I_{11} + \frac{22D_1}{20}I_{12} \qquad +\frac{4D_x}{b}I_{11} - \frac{2bD_1}{15}I_{13}$$

$$-\frac{D_1}{10}I_{10} - \frac{136b^2D_y}{420}I_8 \qquad +\frac{bD_1}{30}I_{10} - \frac{2b^3D_y}{280}I_8 \qquad +\frac{2D_1}{20}I_{13} - \frac{11b^2}{210}I_{14} \qquad -\frac{2bD_1}{15}I_{12} + \frac{2b^3D_y}{210}I_{14}$$

$$+\frac{2D_{xy}}{5}I_9 \qquad\qquad -\frac{2bD_{xy}}{15}I_9 \qquad\qquad -\frac{2D_{xy}}{5}I_{15} \qquad\qquad +\frac{8bD_{xy}}{15}I_{15}$$

$$I_1 = \int_0^a (Y_1)_m(Y_1)_n\,dy, \quad I_2 = \int_0^a (Y_1)_m(Y_1'')_n\,dy, \quad I_3 = \int_0^a (Y_1'')_m(Y_1'')_n\,dy, \quad I_4 = \int_0^a (Y_1'')_m(Y_1'')_n\,dy, \quad I_5 = \int_0^a (Y_1')_m(Y_1')_n\,dy,$$

$$I_6 = \int_0^a (Y_1)_m(Y_2)_n\,dy, \quad I_7 = \int_0^a (Y_1)_m(Y_2'')_n\,dy, \quad I_8 = \int_0^a (Y_1'')_m(Y_2'')_n\,dy, \quad I_9 = \int_0^a (Y_1')_m(Y_2')_n\,dy, \quad I_{10} = \int_0^a (Y_1'')_m(Y_2)_n\,dy,$$

$$I_{11} = \int_0^a (Y_2)_m(Y_2)_n\,dy, \quad I_{12} = \int_0^a (Y_2)_m(Y_2'')_n\,dy, \quad I_{13} = \int_0^a (Y_2'')_m(Y_2)_n\,dy, \quad I_{14} = \int_0^a (Y_2'')_m(Y_2'')_n\,dy, \quad I_{15} = \int_0^a (Y_2')_m(Y_2')_n\,dy$$

Within the range $0<x<b$, both $[C_1]$ and $[C_2]$ will not be equal to zero, and the end conditions of either simply supported or clamped are never satisfied exactly.

For a curved strip, a similar displacement function is written as

$$
w = \sum_{m=1}^{r}\left[\left[\left(1-\frac{3}{4}R^2+\frac{1}{4}R^3\right) \quad b'\left(R-R^2+\frac{1}{4}R^3\right)\right](\Theta_1)_m \right.
$$
$$
\left.\left[\left(\frac{3}{4}R^2-\frac{1}{4}R^3\right) \quad b'\left(\frac{1}{4}R^3+\frac{1}{2}R^2\right)\right](\Theta_2)_m\right]\begin{Bmatrix} w_1 \\ \psi_1 \\ w_2 \\ \psi_2 \end{Bmatrix}_m \qquad (6.2)
$$

in which the variables have the same meaning as was given in Section 2.3, and $(\Theta_i)_m$ is analogous to $(Y_i)_m$.

With the displacement functions established, the formulation of the stiffness and mass matrices follows the standard procedure. For a rectangular strip, the two matrices are listed in Tables 6.1 and 6.2 respectively.

TABLE 6.2. CONSISTENT MASS MATRIX OF A RECTANGULAR MIXED STRIP

|  | $\dfrac{78b}{210}I_1$ | $\dfrac{11b^2}{210}I_1$ | $\dfrac{9b}{70}I_2$ | $\dfrac{-13b^2}{420}I_2$ |
|---|---|---|---|---|
|  | $\dfrac{11b^2}{210}I_1$ | $\dfrac{b^3}{105}I_1$ | $\dfrac{13b^2}{420}I_2$ | $\dfrac{-3b^3}{420}I_2$ |
| $[m]_{mn}=\varrho t$ |  |  |  |  |
|  | $\dfrac{9b}{70}I_4$ | $\dfrac{13b^2}{420}I_4$ | $\dfrac{13b}{35}I_3$ | $\dfrac{-11b^2}{210}I_3$ |
|  | $\dfrac{-13b^2}{420}I_4$ | $\dfrac{-3b^3}{420}I_4$ | $\dfrac{11b^2}{210}I_3$ | $\dfrac{b^3}{105}I_3$ |

$$
I_1=\int_0^a (Y_1)_m(Y_1)_n\,dy, \quad I_2=\int_0^a (Y_1)_m(Y_2)_n\,dy, \quad I_3=\int_0^a (Y_2)_m(Y_2)_n\,dy,
$$
$$
I_4=\int_0^a (Y_2)_m(Y_1)_n\,dy
$$

Two examples are given here to demonstrate the accuracy achievable by the finite strip method. In the first example, the vibrations of a partially clamped square plate (side length equal to $a$) with different clamped lengths are studied. The length of the central clamped portion is equal to $a/6$, $a/4$, and $a/3$ respectively. The three lowest frequencies are given in Table 6.3 and they are compared with other solutions.[2, 3] The

TABLE 6.3. NATURAL FREQUENCIES (RADIAN/SEC) OF PARTIAL CLAMPED PLATES

| Frequencies | $\leftarrow a \rightarrow$ | $\leftarrow a/6$ | $\leftarrow a/4$ | $\leftarrow a/3$ | |
|---|---|---|---|---|---|
| $\omega_1$ | 19·74 | 26·46 | 27·63 (27·31)[2] | 28·94 (28·3)[3] | 36·00 |
| $\omega_2$ | 49·32 | 50·83 | 52·39 | 54·26 | 73·41 |
| $\omega_3$ | 49·34 | 62·11 | 66·24 | 68·07 | 73·41 |

agreement between the results of these methods is excellent. It is o interest to note that frequencies increase with increased clamped length, as is expected.

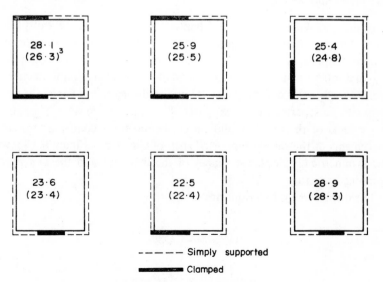

28·1 (26·3)[3]

25·9 (25·5)

25·4 (24·8)

23·6 (23·4)

22·5 (22·4)

28·9 (28·3)

– – – – – Simply supported
━━━━━ Clamped

FIG. 6.2. Mixed boundary plates and their fundamental frequencies.

In the second example, a series of plates with the clamped portions located at different parts of the boundaries are analysed and the fundamental frequencies are compared with those given by Leissa[3] in Fig. 6.2. Again good agreement is obtained for all cases.

## 6.3. PRISMATIC PLATE AND SHELL STRUCTURE WITH FLEXIBLE END SUPPORTS[4]

In Chapter 4 the finite strip method was applied to the analysis of folded plate roofs and box girder bridges supported on rigid end diaphragms. Such structures, however, may be carried on flexible end frames, and the displacements at the supports could have a significant effect on the response of the structures. For simplicity the analysis will be restricted to structures symmetrical with respect to the centre of the span (implying that only symmetrical terms of the series need to be used in the displacement functions), and also flat shell strips will be used so that the membrane action and bending action can be considered separately.

### (a) Membrane action

In the folded plate analysis of Chapter 4 it is assumed that the displacement component $u$ in the plane of the strip is equal to zero at the end support diaphragm. Since the structure is now resting on flexible supports, it is natural to assume that under load the supporting frame members will undergo axial deformations. Thus the displacement functions of a strip can be separated into two parts. The first part consists of parallel movements of the two longitudinal edges and the second part the displacements of the simply supported strip of Chapter 4. Figure 6.3 shows the symmetrically displaced shapes of the first part and the two parts combined together.

The displacement functions are now written as

$$
\left.
\begin{aligned}
u &= \sum_{m=1}^{r} [(1-\bar{x})u_{1m} + (\bar{x})u_{2m}]Y_m, \\
v &= \sum_{m=1}^{r} [(1-\bar{x})v_{1m} + (\bar{x})v_{2m}]Y_m^*,
\end{aligned}
\right\}
\tag{6.3}
$$

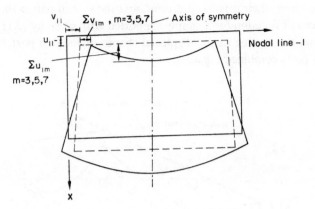

FIG. 6.3. Symmetrically displaced shape of membrane action.

in which

$$Y_m = 1, \quad \sin\frac{\pi y}{a}, \quad \sin\frac{3\pi y}{a}, \ldots, \sin\frac{(m-2)\pi y}{a},$$
$$Y_m^* = \frac{1-2y}{a}, \quad \cos\frac{\pi y}{a}, \quad \cos\frac{3\pi y}{a}, \ldots, \cos\frac{(m-2)\pi y}{a}, \quad (6.4)$$

with $m = 3, 5, 7, \ldots$.

The first term in $Y_m$ and $Y_m^*$ represents the first part of the displacement, while the remaining terms in $Y_m$ and $Y_m^*$ represent the second part. The series is no longer orthogonal, and the stiffness matrix for such a strip will exhibit coupling between the first term and all the other terms of the series.

## (b) Bending action

The same arguments exist for bending action. In Chapter 4 it is assumed that the supports will not yield under any circumstances, while in the present case of flexible supports the frame members will definitely bend when under loading. Therefore it is also possible to separate the displacement function for bending into two parts. The first part causes bending in the transverse $x$ direction only and therefore all the longitudinal lines ($y$ direction) remain straight and horizontal after having been displaced from their original positions, and the second part is simply the displacement function for a simply supported strip given by (2.1). Figure 6.4 shows the symmetrically displaced shapes of the first part and also the two parts combined together.

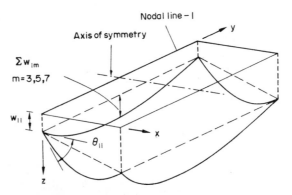

FIG. 6.4. Symmetrically displaced shape of bending action.

The combined displacement function is simply

$$w = \sum_{m=1}^{r} [(1-3\bar{x}^2+2\bar{x}^3)w_{1m} + x(1-2\bar{x}+\bar{x}^2)\theta_{1m} + (3\bar{x}^2-2\bar{x}^3)w_{2m}$$
$$+ x(\bar{x}^2-\bar{x})\theta_{2m}]Y_m \qquad (6.5)$$

in which $Y_m$ has already been defined in (6.4).

Again, the first additional term will couple with all the other terms of the series.


## (c) End beam

The stiffness matrix of a beam in bending and torsion is quite well known and can be found in any textbook on matrix analysis of structures. The neutral axis of the end beam usually does not coincide with the middle surface of the shell strip, and the eccentricity has to be accounted for by performing the transformation given by (2.22a) and (2.22b). From Fig. 6.5

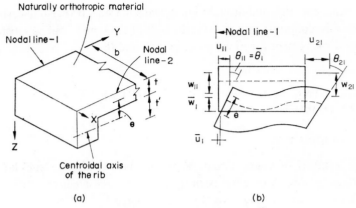

Fig. 6.5. End rib attached to the soffit.

the two sets of displacements can be related to each other through the
following expression:

$$
\begin{Bmatrix} \bar{u}_1 \\ \bar{w}_1 \\ \bar{\theta}_1 \\ \bar{u}_2 \\ \bar{w}_2 \\ \bar{\theta}_2 \end{Bmatrix} =
\begin{bmatrix}
1 & 0 & -e & 0 & 0 & 0 \\
0 & 1 & 0 & 0 & 0 & 0 \\
0 & 0 & 1 & 0 & 0 & 0 \\
0 & 0 & 0 & 1 & 0 & -e \\
0 & 0 & 0 & 0 & 1 & 0 \\
0 & 0 & 0 & 0 & 0 & 1
\end{bmatrix}
\begin{Bmatrix} u_{11} \\ w_{11} \\ \theta_{11} \\ u_{21} \\ w_{21} \\ \theta_{21} \end{Bmatrix}
\tag{6.6}
$$

$$
\{\bar{\delta}\} = [H] \{\delta\} \tag{6.7}
$$

in which $e$ is the eccentricity, $\{\bar{\delta}\}$ and $\{\delta\}$ represent respectively the nodal
displacements of the end beam and the first term displacement parameters
of the strip. The higher term displacement parameters are not related to
the beam displacements since they always specify zero displacements at
the end supports.

After the stiffness matrices for the membrane action and bending
action have been established, they are combined together to form the
stiffness matrix of the flat shell strip. The procedure has already been
described in Section 4.

The example of a cylindrical shell with a canopy is given in reference 4.
The cylindrical shell is supported by rigid diaphragms while the ends of
the canopy are not supported. Unfortunately no comparison was made

with either experimental or other numerical solutions. It is also uncertain whether the condition of $\tau_{xy} = 0$ at the ends of the canopy can be satisfied by the displacement functions given in (6.3) and (6.4).

## 6.4. BENDING AND VIBRATION OF MULTI-LAYER SANDWICH PLATES[5]

### 6.4.1. INTRODUCTION

The analysis of conventional sandwich plates with two stiff layers separated by a weak core has been the topic of extensive investigation in the last few decades, and several reference books[6, 7] have been written on the subject. On the other hand, multi-layer plates with $n$ stiff layers and $n-1$ alternating weak cores have not received much attention. Lungren and Salama[8] have developed a rectangular hybrid element and studied the stability problem of elastic plates, using the multi-layer sandwich plate theory given by Liaw and Little[9] in which the stiff layers are treated as membranes capable of resisting in-plane forces only, and a common shear angle is adopted for all the core layers[†]. However, it has been pointed out by Kao and Ross[10] that the theory is not valid for multi-layer sandwich beams in which the shear strengths of the individual core layers are different. The above statement has been confirmed by Khatua and Cheung[11, 12] for plate problems, using rectangular and triangular finite elements based on displacement models.

The idea of common shear angle is done away with by prescribing arbitrary in-plane displacements in all the stiff layers. While this allows for a more accurate analysis and covers a wider range of problems, the number of DOF for a finite element becomes fairly large because the number of in-plane DOF is now proportional to the number of stiff layers present. Thus the finite strip analysis can be used advantageously because of the significant reduction in the total number of unknowns.

---

[†] In order to assume a common shear angle for all cores, it is necessary to compute an average shear rigidity for core layers as

$$[hG]_{\text{average}} = \frac{1}{n-1} \sum_{j=1}^{n-1} h_j G_j$$

in which $h_j$ is the thickness of the $j^{\text{th}}$ core layer.

### 6.4.2. GENERAL THEORY

*(a) Assumptions*

(1) The stiff layers have relatively high modulus of elasticity and they can withstand both bending and direct forces. However, it has been found that the bending moments in the stiff layers are usually quite small unless fairly thick stiff layers are used, and as a result it is normally safe enough to treat the stiff layers as membranes.

(2) The cores are considered to have zero normal stress stiffness and provide resistance to transverse shear only.

(3) There is no normal strain in the thickness direction, with the result that all stiff layers have the same vertical deflection, although each stiff layer is allowed to have independent in-plane displacements so that each core layer may take up a different shear angle (Fig. 6.6).

(4) The materials can be isotropic, orthotropic, or in general anisotropic.

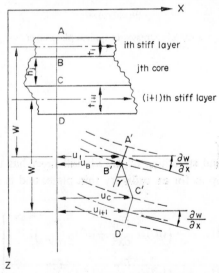

FIG. 6.6. Deformation of stiff layers and core.

*(b) Stress–strain relationships (Fig. 6.7)*

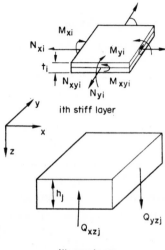

FIG. 6.7. Stresses in typical plate layer.

*1. Stiff layers*

The bending and stretching of the *i*th stiff layer has the same stress-strain relationship as for an ordinary thin plate, and is simply

$$
\left.
\begin{aligned}
M_{xi} &= -\left( (d^b)_{xi}\, \frac{\partial^2 w}{\partial x^2} + (d^b)_{1i}\, \frac{\partial^2 w}{\partial y^2} \right), \\[2mm]
M_{yi} &= -\left( (d^b)_{1i}\, \frac{\partial^2 w}{\partial x^2} + (d^b)_{yi}\, \frac{\partial^2 w}{\partial y^2} \right), \\[2mm]
M_{xyi} &= 2(d^b)_{xyi}\, \frac{\partial^2 w}{\partial x\, \partial y},
\end{aligned}
\right\}
\tag{6.8}
$$

where

$$
\left.\begin{aligned}
(d^b)_{xi} &= \frac{E_{xi}t_i^3}{12(1-\nu_{xi}\nu_{yi})}, \\[2mm]
(d^b)_{yi} &= \frac{E_{yi}t_i^3}{12(1-\nu_{xi}\nu_{yi})}, \\[2mm]
(d^b)_{1i} &= \nu_{yi}(d^b)_{xi} = \nu_{xi}(d^b)_{yi}, \\[2mm]
(d^b)_{xyi} &= \frac{E_{xyi}t_i^3}{12},
\end{aligned}\right\}
\tag{6.9}
$$

and the suffix $i$ denotes the $i$th layer.

### 2. Shear deformation in the jth core layer

Only shear deformation needs to be considered for the case of a weak core layer since all the other stress (strain) components are assumed to be zero. The relationships between shear force, shear strain, and displacements are somewhat more complicated and are worthy of some discussion.

From Fig. 6.6 it is seen that the shear strain in the $xz$ plane of the $j$th core is represented by the angle $\gamma$, which is in turn made up of two parts; one part is due to bending of the plate and is equal to the slope $\partial w/\partial x$, while the other part is due to shearing of the $j$th core layer and is equal to $(u_C - u_B)/h_j$, i.e.

$$
\gamma_{xzj} = \frac{u_C - u_B}{h_j} + \frac{\partial w}{\partial x}.
\tag{6.10}
$$

It is now necessary to relate $u_C$ and $u_B$ to the nodal displacement parameters $u_i$ and $u_{i+1}$. Again, from Fig. 6.6,

$$
\left.\begin{aligned}
u_C &= u_{i+1} + \frac{t_{i+1}}{2}\frac{\partial w}{\partial x}, \\[2mm]
u_B &= u_i - \frac{t_i}{2}\frac{\partial w}{\partial x}.
\end{aligned}\right\}
\tag{6.11}
$$

Therefore, substituting (6.11) (6.10),

$$
\gamma_{xzj} = \frac{1}{h_j}\left(\left(u_{i+1} + \frac{t_{i+1}}{2}\frac{\partial w}{\partial x}\right) - \left(u_i - \frac{t_i}{2}\frac{\partial w}{\partial x}\right)\right) + \frac{\partial w}{\partial x}.
\tag{6.12}
$$

Simplifying (6.12),

$$\gamma_{xzj} = \frac{C_j}{h_j}\left(\frac{u_{i+1}-u_j}{C_j}+\frac{\partial w}{\partial x}\right), \tag{6.13}$$

where

$$C_j = h_j+\frac{1}{2}(t_{i+1}+t_i). \tag{6.14}$$

The shear force $Q_{xzj}$ of the $j$th core layer can now be expressed in terms of the shear strain as

$$Q_{xzj} = (G_{xzj}h_j)\gamma_{xzj} = (d^s)_{xzj}\gamma_{xzj}. \tag{6.15}$$

Similar expressions can be derived for the shearing in the $yz$ plane, and we have

$$\gamma_{yzj} = \frac{C_j}{h_j}\left(\frac{v_{i+1}-v_i}{C_j}+\frac{\partial w}{\partial y}\right) \tag{6.16}$$

and

$$Q_{yzj} = (G_{yzj}h_j)\gamma_{yzj} = (d^s)_{yzj}\gamma_{yzj}. \tag{6.17}$$

### 6.4.3. FORMULATION OF STIFFNESS AND MASS MATRICES

A lower order multi-layer sandwich strip will be developed herein. There are two in-plane displacements $u_i$ and $v_i$ (linear polynomial of $x$) and two out-of-plane displacement parameters $w_i$ and $\theta_i$ (cubic polynomial of $x$) for each individual stiff layer at nodal line $i$. Due to assumption (2) $w_i$ and $\theta_i$ will be the same for all stiff layers. Thus the number of unknown parameters for a nodal line of a strip with $n$ stiff layers will be $(2+2n)$ times the number of terms used in the series part $(Y_m)$ of the displacement function. Note that in static analysis the end condition at a support should either be simply supported or clamped, while for vibration analysis all six types of end conditions listed in Section 1.3.1 can be included.

The deformation of a multi-layer sandwich strip can be viewed as the combined result of three separate processes, i.e. the bending of the stiff layers, the stretching of the stiff layers, and the shearing of the cores, which couples the different stiff layers together. The bending of the stiff layers is directly related to the sum of the flexural rigidities of the individual since the bending displacement parameters are the same for all the layers. Schematically the formulation of the stiffness matrix is as shown in Fig. 6.8. The submatrix $[S_{ijk}^b]$ is the total bending stiffness of all the stiff layers and is related to the nodal displacement parameters $w_1$ and $\theta_1$ or $w_2$ and $\theta_2$, and $[S_{ijk}^p]$ is the in-plane stiffness of each individual

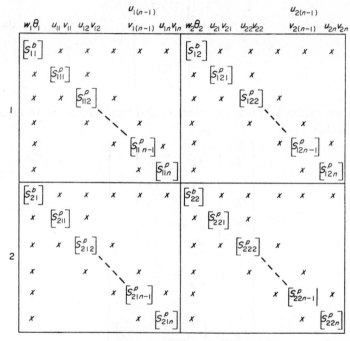

FIG. 6.8. Schematic diagram of stiffness matrix.

stiff layer and is related to the displacement parameters $u_{1k}$, $v_{1k}$ or $u_{2k}$, $v_{2k}$ (here the suffix $k$ refers to the $k$th stiff layer and not in the usual context to the $k$th terms of the series), in which $i$ and $j$ (1 and 2) are the nodal line numbers. These submatrices can be obtained by the established procedures for finite strips. The effect of the shearing of the cores is to couple the different stiff layers together or, in other words, to fill up some of the empty space of the schematic diagram in Fig. 6.8.

Listing the nodal displacement parameters as

$$\{\delta\} = [w_1\theta_1 u_{11} v_{11} u_{12} v_{12} \dots u_{1n} v_{1n} w_2\theta_2 u_{21} v_{21} u_{22} v_{22} \dots u_{2n} v_{2n}]^T \quad (6.18)$$

and the shear strains for the $n-1$ core layer as

$$\{\varepsilon\} = \begin{Bmatrix} \gamma_{xz1} \\ \gamma_{yz1} \\ \gamma_{xz2} \\ \gamma_{yz2} \\ \vdots \\ \gamma_{xz\,n-1} \\ \gamma_{yz\,n-1} \end{Bmatrix} \quad (6.19)$$

the strain matrix $[B^s]$ is easily established through the use of (6.13) and (6.16) .Thus

$$\{\varepsilon^s\} = \begin{Bmatrix} \dfrac{C_1}{h_1}\left(\dfrac{u_2-u_1}{C_1}+\dfrac{\partial w}{\partial x}\right) \\[2ex] \dfrac{C_1}{h_1}\left(\dfrac{v_2-v_1}{C_1}+\dfrac{\partial w}{\partial y}\right) \\[2ex] \dfrac{C_2}{h_2}\left(\dfrac{u_3-u_2}{C_2}+\dfrac{\partial w}{\partial x}\right) \\[2ex] \dfrac{C_2}{h_2}\left(\dfrac{v_3-v_2}{C_2}+\dfrac{\partial w}{\partial y}\right) \\[2ex] \dots \\[2ex] \dfrac{C_{n-1}}{h_{n-1}}\left(\dfrac{u_n-u_{n-1}}{C_{n-1}}+\dfrac{\partial w}{\partial x}\right) \\[2ex] \dfrac{C_{n-1}}{h_{n-1}}\left(\dfrac{v_n-v_{n-1}}{C_{n-1}}+\dfrac{\partial w}{\partial y}\right) \end{Bmatrix} \quad (6.20)$$

$$= [B^s]\{\delta\}.$$

The stiffness matrix due to the shearing of the core can now be readily obtained through the familiar relationship of

$$[S^s] = \int [B^s]^T [D^s] [B^s] \, dx \, dy, \quad (6.21)$$

where $[D^s]$ is a diagonal matrix of the shear constants ($d^s$) of the cores.

The mass matrix is obtained in the usual manner through the use of (5.5), and the arrangement of the submatrices is also depicted by Fig. (6.8) in which $[S_{ij}^b]$ and $[S_{ijk}^p]$ should be replaced by the bending mass matrix $[M_{ij}^b]^e$ and in-plane mass matrix $[M_{ijk}^p]^e$ respectively. For the $[M_{ij}^b]^e$ the overall mass distribution of the plate, obtained by summing up the mass of every layer including cores, is used. For $[M_{ijk}^p]^e$ the mass of half of the thickness of the adjacent core layers and the mass of the $k$th stiff layer are included. There is no coupling effect between the in-plane and out-of-plane mass matrices, and therefore all the crosses in Fig. 6.8 are equal to zero.

### 6.4.4. NUMERICAL EXAMPLES

#### (a) Bending problems

Three- and five-layer isotropic and orthotropic sandwich plates with various boundary conditions were analysed. Only one half of the plate with five strips was used wherever symmetry existed. The results are listed in Tables 6.4 and 6.5 along with available published results, and good agreement is observed for all cases. In Table 6.6 central deflections and moments and maximum edge moments are given for several boundary conditions, and serve to demonstrate the versatility of the method.

#### (b) Vibration problems

Natural frequencies for rectangular five-layer sandwich plates were obtained by this method and compared with published results (Table 6.7). The advantage of the finite strip method over the conventional finite element method is more obvious for such eigenvalue problems since a much smaller matrix is resulted from the finite strip method formulation.

#### (c) Shear angles of cores

An isotropic five-layer sandwich plate with shear constants of the two cores having a ratio of ten was analysed. The dimensions and properties of the plate are the same as shown in Table 6.5 except that $G_{xz1} = G_{yz1} =$

TABLE 6.4. THREE-LAYER SIMPLY SUPPORTED ISOTROPIC AND ORTHOTROPIC SANDWICH PLATE UNDER UNIFORM LOAD

| Terms | Central deflection (m) | | Central bending moment $M_x$ (kN-m/m) | | Central bending moment $M_y$ (kN-m/m) | |
|---|---|---|---|---|---|---|
| | Isotropic | Orthotropic | Isotropic | Orthotropic | Isotropic | Orthotropic |
| 1 | 0.0007642 | 0.0012689 | 4.962 | 7.677 | 5.229 | 3.265 |
| 3 | −0.0000248 | −0.0000595 | −0.157 | −0.280 | −0.457 | −0.420 |
| 5 | 0.0000047 | 0.0000117 | 0.031 | 0.040 | 0.012 | 0.100 |
| 7 | −0.0000016 | −0.0000041 | −0.011 | −0.012 | −0.037 | −0.037 |
| $\Sigma$ | 0.0007435 | 0.001217 | 4.825 | 7.425 | 4.838 | 2.908 |
| Series solution[9] | 0.0007395 | — | 4.79 | — | 4.79 | — |
| Finite element[11] | 0.0007361 | 0.001213 | 4.7789 | 7.4433 | 4.7789 | — |

Note: $A = 10$ m; $B = 10$ m; $t_1 = t_2 = 0.028$ m; $h_1 = 0.75$ m; $q = 1$ kN/m$^2$.

Isotropic case:

$E_{x1} = E_{x2} = E_{y1} = E_{y2} = 10^7$ kN/m$^2$; $G_{yz1} = G_{yz1} = 3 \times 10^4$ kN/m$^2$; $\nu_{x1} = \nu_{y1} = \nu_{z2} = \nu_{y2} = 0.3$.

Orthotropic case:

$E_{x1} = E_{x2} = 10^7$ kN/m$^2$; $E_{y1} = E_{y2} = 4 \times 10^6$ kN/m$^2$; $E_{xy1} = E_{xy2} = 1.875 \times 10^6$ kN/m$^2$.

$\nu_{x1} = \nu_{x2} = 0.3$; $\nu_{y1} = \nu_{y2} = 0.12$; $G_{zz1} = 3 \times 10^4$ kN/m$^2$; $G_{yz1} = 0.2 \times 10^4$ kN/m$^2$.

TABLE 6.5. FIVE-LAYER SIMPLY SUPPORTED ISOTROPIC AND ORTHOTROPIC SANDWICH PLATE UNDER UNIFORM LOAD

| Terms | Central deflection (m) | | Central bending moment $M_x$ (kN-m/m) | | Central bending moment $M_y$ (kN-m/m) | |
|---|---|---|---|---|---|---|
| | Isotropic | Orthotropic | Isotropic | Orthotropic | Isotropic | Orthotropic |
| 1 | 0.0008264 | 0.0013884 | 4.964 | 7.603 | 5.231 | 3.319 |
| 3 | −0.0000243 | −0.0000586 | −0.157 | −0.266 | −0.457 | −0.421 |
| 5 | 0.0000044 | 0.0000111 | 0.031 | 0.039 | 0.103 | 0.101 |
| 7 | −0.0000015 | −0.0000038 | −0.012 | −0.012 | −0.037 | −0.037 |
| $\Sigma$ | 0.0008050 | 0.0013371 | 4.826 | 7.365 | 4.838 | 2.962 |
| Finite element[11] | 0.0007968 | 0.001332 | 4.7740 | 7.4012 | 4.7740 | — |

Note: $A = 10$ m; $B = 10$ m; $t_1 = t_2 = t_3 = 0.02$ m; $h_1 = h_2 = 0.4$ m; $q = $ ' kN/m²

Elastic properties of materials are the same as in Table 6.4.

TABLE 6.6. THREE-LAYER ISOTROPIC SANDWICH PLATE UNDER UNIFORM LOAD

| Boundary conditions | Central deflection (m) | Central moments (kN-m/m) | | Edge moment $M_y$ (kN-m/m) |
|---|---|---|---|---|
| | | $M_x$ | $M_y$ | |
| S–c–s–s | 0·0006036 | 3·828 | 3·709 | –6·409 |
| C–c–s–s | 0·0005449 | 3·437 | 3·185 | –5·105 |
| C–c–c–s | 0·0004647 | 2·66 | 2·74 | –4·99 |
| C–c–c–c | 0·0004244 | 2·360 | 2·368 | –4·361 |
| C–s–c–s | 0·000529 | 3·13 | 3·13 | –5·745 |

Note: Dimensions and elastic properties are the same as Table 6.4.

$= 30\,000$ kN/m$^2$ and $G_{xz2} = G_{yz2} = 3000$ kN/m$^2$. The in-plane displacements of the stiff layers in the $x$ direction, $u$, at the quarter point of the plate are plotted and shown in Fig. 6.9. It can easily be seen that the shear angles of the two cores differ significantly. This shows that the increase in displacement parameters ($u_i$ and $v_i$ for each individual stiff layer) to allow for this effect is justified, and that the assumption of common shear angle for all cores can be quite erroneous for certain cases.

TABLE 6.7. NATURAL FREQUENCIES (CPS) OF A FIVE-LAYER SIMPLY SUPPORTED SANDWICH PLATE

| Modal numbers | $m'$ | 1 | 2 | 1 | 3 | 2 | 3 | 4 | 1 | 2 | 4 |
|---|---|---|---|---|---|---|---|---|---|---|---|
| | $n'$ | 1 | 1 | 2 | 1 | 2 | 2 | 1 | 3 | 3 | 2 |
| Finite element Ref. 11 | | 19 | 38 | 60 | 69 | 78 | 109 | 115 | 128 | 145 | 153 |
| Finite strip | | 20 | 38 | 60 | 68 | 78 | 107 | 109 | 127 | 144 | 148 |

Note:

$n = 3$; $A = 1.83$ m; $B = 1.22$ m; $t_1 = t_2 = t_3 = 0.00028$ m; $h = 0.00318$ m;

$E_{x1} = E_{x2} = E_{x3} = E_{y1} = E_{y2} = E_{y3} = 6.89 \times 10^7$ kN/m$^2$; $G_{xz1} = G_{xz2} =$

$= 134.45 \times 10^3$ kN/m$^2$; $G_{yz1} = G_{yz2} = 51.71 \times 10^3$ kN/m$^2$; $\nu_{x1} = \nu_{y1} = \nu_{x2} =$

$= \nu_{y2} = \nu_{x3} = \nu_{y3} = 0.33$; $\varrho_{s1} = \varrho_{s2} = \varrho_{s3} = 2.767$ kN/s$^2$/m$^4$.

$\varrho_{c1} = \varrho_{c2} = 0.123$ kN/s$^2$/m$^4$.

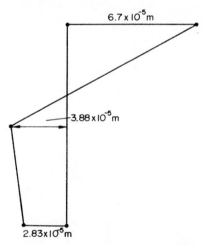

FIG. 6.9. Displacement $u$ of stiff layers.

REFERENCES

1. M. S. CHEUNG, Finite strip analysis of structures, PhD thesis, University of Calgary, 1971.
2. M. KURATA and H. OKAMURA, Natural vibration of partially clamped plates, Am. Soc. Civ. Engrs 89, EM (June 1963).

3. A. W. LEISSA, Free vibration of elastic plates, *AIAA 7th Aerospace Sciences Meeting, New York, January 1969.*

4. G. H. SIDDIQI and C. V. GIRIJA–VALLABHAN, Extended finite strip method for prismatic plate and shell structures, *IASS Pacific Symposium on Hydromechanically Loaded Shells, University of Hawaii, October 1971.*

5. H. C. CHAN and Y. K. CHEUNG, Static and dynamic analysis of multi-layered sandwich plates, *Int. J. Mech. Sci.* **14,** 399–406 (1972).

6. H. G. ALLEN, *Analysis and Design of Structural Sandwich Panels,* Pergamon Press, London, 1969.

7. F. J. PLANTEMA, *Sandwich Construction,* John Wiley, New York, 1966.

8. H. R. LUNGREN and A. E. SALAMA, Buckling of multilayer plates by finite elements, *Am. Soc. Civ. Engrs* **97,** EM2, 477–94 (April 1971).

9. B. D. LIAW and R. W. LITTLE, Theory of bending multilayer sandwich plates, *AIAA Jl* **5,** (2) 301–4 (February 1967).

10. J. S. KAO and R. J. Ross, Bending of multilayer sandwich beams, *AIAA Jl* **6** (8) 961 (August 1968).

11. T. P. KHATUA and Y. K. CHEUNG, Bending and vibration of multilayer sandwich beams and plates, *Int. J. Num. Meth. Eng.* **6** (1973).

12. T. P. KHATUA and Y. K. CHEUNG, Triangular element for multilayer sandwich plates, *Am. Soc. Civ. Engrs* **98,** EM5, 1225–38 (October 1972).

# CHAPTER 7

# *Finite layer method and finite prism method*

## 7.1. INTRODUCTION

In the previous chapters the development of the finite strip method and application to two-dimensional problems have been discussed in detail. Here attention will be paid to the analysis of three-dimensional problems.

It was mentioned briefly in Chapter 1 that it is possible to reduce a three-dimensional problem to a one-dimensional one by assuming displacement functions which take the form of

$$f = \sum_{m=1}^{r} \sum_{n=1}^{t} f_{mn}(z) X_m Y_n, \tag{1.1c}$$

thus forming the basis of the finite layer method. In this method, a thick plate is imagined to be divided into a number of horizontal layers with each being supported along the four vertical sides (Fig. 7.1A), while a cylinder is divided into a number of concentric cylindrical layers (Fig. 7.1B). Since each layer can be assigned individual material properties and thickness, the method is ideally suited for the analysis of thick, layered plates and shells.

Similarly, a three-dimensional problem can be reduced to a two-dimensional one by writing the displacement function as

$$f = \sum_{m=1}^{r} f_m(x, z) Y_m. \tag{1.1b}$$

The above displacement function can then be used to formulate the stiffness matrix of a finite prism. This method is best suited to structures with variable thicknesses or with holes in the cross-section, such as voided slabs or thick-walled box girder bridges. The investigation of

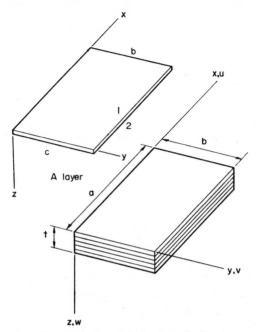

FIG. 7.1A. Finite layer idealization of thick plate.

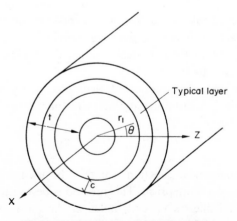

FIG. 7.1B. Layer of a thick cylinder.

local stresses at web and flange junctions, which normally would not be justified because of the high cost of a full three-dimensional finite element analysis, can now be carried out at moderate cost.

## 7.2. FINITE LAYER METHOD

### 7.2.1. RECTANGULAR PRISM LAYER

The rectangular prism layer shown in Fig. 7.1A has two nodal surfaces and a suitable set of displacement functions is selected as

$$
\left.
\begin{aligned}
u &= \sum_{m=1}^{r} \sum_{n=1}^{t} [(1-\bar{z})u_{1mn}+\bar{z}u_{2mn}]\, X'_m(x)\, Y_n(y), \\[2mm]
v &= \sum_{m=1}^{r} \sum_{n=1}^{t} [(1-\bar{z})v_{1mn}+\bar{z}v_{2mn}]\, X_m(x)\, Y'_n(y), \\[2mm]
w &= \sum_{m=1}^{r} \sum_{n=1}^{t} [(1-\bar{z})w_{1mn}+\bar{z}w_{2mn}]\, X_m(x)\, Y_n(y),
\end{aligned}
\right\}
\qquad (7.1)
$$

in which $u_{1mn}$, $v_{1mn}$, $w_{1mn}$, ..., etc., refer to the displacement parameters, and $X_m(x)$, $X'_m(x)$, $Y_n(y)$, $Y'_n(y)$ are the basic functions and their first derivatives. Note that $X'_m(x)$ and $Y'_n(y)$ are incorporated in the expressions for $u$ and $v$ respectively because of the relationship $u = A(\partial w/\partial x)$ and $v = B(\partial w/\partial y)$ in linear plate theory, and they are used here as an approximation for each individual layer. For two-dimensional problems the validity of such an assumption has been amply demonstrated in the examples of Chapter 2.

The displacement in matrix form is

$$
\{f\} = \begin{Bmatrix} u \\ v \\ w \end{Bmatrix} = \sum_{m=1}^{r} \sum_{n=1}^{t} [N]_{mn}\,\{\delta\}_{mn} = [N]\,\{\delta\} \qquad (7.2)
$$

in which $\{\delta\}_{mn} = [u_{1mn},\ v_{1mn},\ w_{1mn},\ u_{2mn},\ v_{2mn},\ w_{2mn}]^T$.

The strain matrix of a three-dimensional solid is

$$\{\varepsilon\} = \begin{Bmatrix} \varepsilon_x \\ \varepsilon_y \\ \varepsilon_z \\ \gamma_{xy} \\ \gamma_{yz} \\ \gamma_{zx} \end{Bmatrix} = \begin{Bmatrix} \dfrac{\partial u}{\partial x} \\ \dfrac{\partial v}{\partial y} \\ \dfrac{\partial w}{\partial z} \\ \dfrac{\partial u}{\partial y} + \dfrac{\partial v}{\partial x} \\ \dfrac{\partial v}{\partial z} + \dfrac{\partial w}{\partial y} \\ \dfrac{\partial w}{\partial x} + \dfrac{\partial u}{\partial z} \end{Bmatrix} \tag{7.3a}$$

or

$$\{\varepsilon\} = \sum_{m=1}^{r} \sum_{n=1}^{t} [B]_{mn} \{\delta\}_{mn} = [B]\{\delta\}. \tag{7.3b}$$

The stress–strain relationships are

$$\{\sigma\} = [D]\{\varepsilon\}$$

$$= [D] \sum_{m=1}^{r} \sum_{n=1}^{t} [B]_{mn} \{\delta\}_{mn}$$

$$= [D][B]\{\delta\} \tag{7.4a}$$

in which $\{\sigma\}$ represents the stresses corresponding to the strains $\{\varepsilon\}$ and is given by

$$\{\sigma\} = \begin{Bmatrix} \sigma_x \\ \sigma_y \\ \sigma_z \\ \tau_{xy} \\ \tau_{yz} \\ \tau_{zx} \end{Bmatrix}. \tag{7.4b}$$

$[D]$ is the elasticity matrix of the material for the particular layer under consideration which can be isotropic, orthotropic, or general anisotropic.
The standard formula for the stiffness matrix is given by (1.29a) as

$$[S] = \int [B]^T [D][B] \, d(\text{vol.})$$

The expanded form of the strain matrix is

$$[B] = [[B]_{11}[B]_{12} \ldots [B]_{1t} \quad [B]_{21}[B]_{22} \ldots [B]_{2t} \ldots [B]_{r1}[B]_{r2} \ldots [B]_{rt}]. \quad (7.5)$$

Therefore the stiffness matrix for a layer is

$$[S] = \int [B]^T [D] [B] \, d \, (\text{vol.})$$

$$= \int_0^a \int_0^b \int_0^c$$

$$\begin{bmatrix}
[B]_{11}^T [D] [B]_{11} & [B]_{11}^T [D] [B]_{12} & \ldots & [B]_{11}^T [D] [B]_{1t} & [B]_{11}^T [D] [B]_{21} & [B]_{11}^T [D] [B]_{22} & \ldots & [B]_{11}^T [D] [B]_{2t} & \ldots & [B]_{11}^T [D] [B]_{rt} \\
[B]_{12}^T [D] [B]_{11} & [B]_{12}^T [D] [B]_{12} & \ldots & [B]_{12}^T [D] [B]_{1t} & [B]_{12}^T [D] [B]_{21} & [B]_{12}^T [D] [B]_{22} & \ldots & [B]_{12}^T [D] [B]_{2t} & \ldots & [B]_{12}^T [D] [B]_{rt} \\
\ldots & \ldots & & \ldots & \ldots & \ldots & & \ldots & & \ldots \\
[B]_{1t}^T [D] [B]_{11} & [B]_{1t}^T [D] [B]_{12} & \ldots & [B]_{1t}^T [D] [B]_{1t} & [B]_{1t}^T [D] [B]_{21} & [B]_{1t}^T [D] [B]_{22} & \ldots & [B]_{1t}^T [D] [B]_{2t} & \ldots & [B]_{1t}^T [D] [B]_{rt} \\
[B]_{21}^T [D] [B]_{11} & [B]_{21}^T [D] [B]_{12} & \ldots & [B]_{21}^T [D] [B]_{1t} & [B]_{21}^T [D] [B]_{21} & [B]_{21}^T [D] [B]_{22} & \ldots & [B]_{21}^T [D] [B]_{2t} & \ldots & [B]_{21}^T [D] [B]_{rt} \\
[B]_{22}^T [D] [B]_{11} & [B]_{22}^T [D] [B]_{12} & \ldots & [B]_{22}^T [D] [B]_{1t} & [B]_{22}^T [D] [B]_{21} & [B]_{22}^T [D] [B]_{22} & \ldots & [B]_{22}^T [D] [B]_{2t} & \ldots & [B]_{22}^T [D] [B]_{rt} \\
\ldots & \ldots & & \ldots & \ldots & \ldots & & \ldots & & \ldots \\
[B]_{rt}^T [D] [B]_{11} & [B]_{rt}^T [D] [B]_{12} & \ldots & [B]_{rt}^T [D] [B]_{1t} & [B]_{rt}^T [D] [B]_{21} & [B]_{rt}^T [D] [B]_{22} & \ldots & [B]_{rt}^T [D] [B]_{2t} & \ldots & [B]_{rt}^T [D] [B]_{rt}
\end{bmatrix} dx \, dy \, dz \quad (7.6)$$

The consistent mass matrix for a layer is developed in a similar way. From (5.5) we have

$$[M]^e = \int \rho [N]^T [N] \, d\,(\text{vol.})$$

$$= \int_0^a \int_0^b \int_0^c
\begin{bmatrix}
[N]_{11}^T [N]_{11} & [N]_{11}^T [N]_{12} & \cdots & [N]_{11}^T [N]_{1t} & [N]_{11}^T [N]_{21} & [N]_{11}^T [N]_{22} & \cdots & [N]_{11}^T [N]_{2t} & \cdots & [N]_{11}^T [N]_{rt} \\
[N]_{12}^T [N]_{11} & [N]_{12}^T [N]_{12} & \cdots & [N]_{12}^T [N]_{1t} & [N]_{12}^T [N]_{21} & [N]_{12}^T [N]_{22} & \cdots & [N]_{12}^T [N]_{2t} & \cdots & [N]_{12}^T [N]_{rt} \\
\cdots & \cdots & \cdots & \cdots & \cdots & \cdots & \cdots & \cdots & \cdots & \cdots \\
[N]_{1t}^T [N]_{11} & [N]_{1t}^T [N]_{12} & \cdots & [N]_{1t}^T [N]_{1t} & [N]_{1t}^T [N]_{21} & [N]_{1t}^T [N]_{22} & \cdots & [N]_{1t}^T [N]_{2t} & \cdots & [N]_{1t}^T [N]_{rt} \\
[N]_{21}^T [N]_{11} & [N]_{21}^T [N]_{12} & \cdots & [N]_{21}^T [N]_{1t} & [N]_{21}^T [N]_{21} & [N]_{21}^T [N]_{21} & \cdots & [N]_{21}^T [N]_{2t} & \cdots & [N]_{21}^T [N]_{rt} \\
[N]_{22}^T [N]_{11} & [N]_{22}^T [N]_{12} & \cdots & [N]_{22}^T [N]_{1t} & [N]_{22}^T [N]_{21} & [N]_{22}^T [N]_{22} & \cdots & [N]_{22}^T [N]_{2t} & \cdots & [N]_{22}^T [N]_{rt} \\
\cdots & \cdots & \cdots & \cdots & \cdots & \cdots & \cdots & \cdots & \cdots & \cdots \\
[N]_{rt}^T [N]_{11} & [N]_{rt}^T [N]_{12} & \cdots & [N]_{rt}^T [N]_{1t} & [N]_{rt}^T [N]_{21} & [N]_{rt}^T [N]_{22} & \cdots & [N]_{rt}^T [N]_{2t} & \cdots & [N]_{rt}^T [N]_{rt}
\end{bmatrix}
dx \, dy \, dz \quad (7.7)$$

Note that the final form of the stiffness matrix and mass matrix [(7.6) and (7.7)] are somewhat different from the ones given by (1.29a) and (5.5) because of the double summation used in the displacement functions. The assembly of this type of stiffness and mass matrices will be discussed in Chapter 8.

Only the static analysis of a simply supported plate has been attempted, and for such a problem we have

$$\int_0^a \int_0^b \int_0^c [B]_{mn}^T [D] [B]_{rs} \, dx \, dy \, dz = 0 \quad \text{for} \quad mn \neq rs, \qquad (7.8)$$

i.e. all the off-diagonal submatrices are zero and the terms of the series are now decoupled.

For vibration problems, however, all combinations of end conditions are allowable.

To demonstrate the application of the finite layer method to the solution of thick plate problems, the following two examples have been considered.

The first example is on the static analysis of a simply supported plate[1] with uniformly distributed load $q$ acting at the top surface. The results for various ratios of $b/a$ and $t/a$ are shown in Table 7.1, using twenty-five terms of the series ($m = 1, 3, 5, 7, 9$ and $n = 1, 3, 5, 7, 9$). The plate is

TABLE 7.1. STRESSES AND DEFLECTIONS AT CENTER OF RECTANGULAR PLATES UNDER UNIFORMLY DISTRIBUTED LOAD ($a = 10$, $E = 1$, $v = 0.15$, $q = 1$)[a]

| $b/a$ | $t/a$ | 0.10 | | 0.25 | | 0.50 | |
|---|---|---|---|---|---|---|---|
| | | Top | Bottom | Top | Bottom | Top | Bottom |
| 1.0 | $w$ | 491.9639 | 491.4200 | 37.7139 | 36.4321 | 9.0208 | 6.0001 |
| | $\sigma_x$ | −26.0287 | 25.9095 | −4.2806 | 4.1699 | −1.2427 | 1.0938 |
| | $\sigma_y$ | −26.0287 | 25.9095 | −4.2806 | 4.1699 | −1.2427 | 1.0938 |
| 1.5 | $w$ | 924.2100 | 923.6643 | 67.4901 | 66.1891 | 13.0846 | 11.0972 |
| | $\sigma_x$ | −46.7853 | 46.7066 | −7.6061 | −7.5305 | −2.0725 | 1.9684 |
| | $\sigma_y$ | −24.1373 | 23.9917 | −4.0081 | 3.8476 | −1.1594 | 0.9990 |
| | $w$ | 1207.1999 | 1206.6533 | 86.6447 | 85.3310 | 16.5533 | 13.9514 |
| | $\sigma_x$ | −60.0407 | 59.9828 | −9.7299 | 9.6724 | −2.6056 | 2.5186 |
| | $\sigma_y$ | −20.0641 | 19.9060 | −3.3642 | 3.1857 | −1.0032 | 0.8134 |

[a]Tension = +ve, compression = −ve

divided into three equal layers for $t/a$ ratios of 0.10 and 0.25, and into five equal layers when $t/a$ is 0.50. Note that for $t/a = 0.10$ the results are close to those given by the thin-plate theory, and the deflections and stresses are almost equal at upper and lower extreme fibres, although in the finite layer method the thin plate is treated as a three-dimensional solid. For higher $t/a$ ratios, the thin plate theory is no longer applicable, and both the stresses and deflections at the extreme fibres show marked differences in magnitudes.

Note that $\varepsilon_z$ is constant through the thickness of each layer and consequently for non-zero Poisson ratios stress jumps can be expected at the

TABLE 7.2. NATURAL FREQUENCIES $f(Hz)$ FOR VARIOUS ISOTROPIC HOMOGENEOUS PLATES

| Cases | Frequency number | Thin plate theory, $f/h$ | $h = 0.05$ | $h = 0.10$ | $h = 0.20$ |
|---|---|---|---|---|---|
| (clamped all edges) | 1 | 1.7402 | 0.0887 | 0.1662 | 0.2744 |
| | 2 | 3.5518 | 0.1777 | 0.3220 | 0.4981 |
| | 3 | 3.5518 | 0.1777 | 0.3220 | 0.4981 |
| | 4 | 5.2402 | 0.2576 | 0.4547 | 0.6741 |
| | 5 | 6.3813 | 0.3043 | 0.5166 | 0.7245 |
| | 1 | 1.1671 | 0.0597 | 0.1134 | 0.1930 |
| | 2 | 1.9525 | 0.0975 | 0.1819 | 0.2981 |
| | 3 | 3.0812 | 0.1550 | 0.2834 | 0.4451 |
| | 4 | 3.7245 | 0.1833 | 0.3334 | 0.5207 |
| | 5 | 3.9550 | 0.1937 | 0.3491 | 0.5346 |
| | 1 | 1.3100 | 0.4153 | 0.1268 | 0.2222 |
| | 2 | 2.9377 | 0.1454 | 0.2688 | 0.4326 |
| | 3 | 2.9377 | 0.1454 | 0.2688 | 0.4326 |
| | 4 | 4.4853 | 0.2190 | 0.3937 | 0.6024 |
| | 5 | 5.5475 | 0.2671 | 0.4702 | 0.6949 |
| | 1 | 1.3972 | 0.0709 | 0.1344 | 0.2278 |
| | 2 | 2.6520 | 0.1305 | 0.2430 | 0.3984 |
| | 3 | 3.3402 | 0.1677 | 0.3056 | 0.4775 |
| | 4 | 4.5675 | 0.2239 | 0.4013 | 0.6110 |
| | 5 | 4.9538 | 0.2374 | 0.4260 | 0.6513 |
| | 1 | 1.1418 | 0.0574 | 0.1109 | 0.1977 |
| | 2 | 2.5044 | 0.1228 | 0.2309 | 0.3859 |
| | 3 | 2.8250 | 0.1403 | 0.2602 | 0.4214 |
| | 4 | 4.1548 | 0.2025 | 0.3675 | 0.5730 |
| | 5 | 4.8570 | 0.2327 | 0.4193 | 0.6452 |

Data: 5 identical layers; $m = 1, 2, 3$; $n = 1, 2, 3$; $a = b = 1.0$; $E = 1.0$; $\nu = 0.30$; $\varrho = 1.0$. Simply supported -------- ; free ——— ; clamped //////////.

interfaces of adjoining layers for the normal stress component $\sigma_z$. Therefore the stresses should be averaged at the interface or plotted at the mid-height of layers to obtain a smooth stress distribution.

The second problem concerns the free vibration of isotropic plates[2] with various types of boundary conditions for three different thickness–span ratios. Each plate was analysed with five identical layers and three terms of the double series. Thus, in each case, 162 equations were involved in the computation of the frequencies. Only results for the first five lowest flexural frequencies are shown in Table 7.2. It is observed that the frequencies for the smallest thickness–span ratio are close to those given by Warburton[3] for thin plates, while the frequencies for higher thickness–span ratios tend to be of lower values.

A point worth mentioning is that in comparison with the exact values the thin plate theory tends to overestimate the frequencies and the error increases with higher modes of vibration.

### 7.2.2. CYLINDRICAL LAYER (FIG. 7.1B)

Two types of cylindrical layers have been developed. The first is a higher order layer (HO3) with an additional internal nodal surface developed by Nelson et al.[4] for the study of free vibration of infinite cylinders. The second is a lower order layer (LO2) used by Cheung and Wu[5] for the determination of the frequencies of thick, layered cylinders with any combinations of boundary conditions at the axial ends of the cylinder.

#### (a) Lower order cylindrical layer (LO2)

For a typical cylindrical layer with nodal surfaces 1 and 2, the thickness is designated by $c$. The layer is referred to a right-hand screw coordinate system $(x, \theta, z)$ with $x$ being taken along the axial direction, $\theta$ the circumferential direction, and $z$ the radial direction, which is positive when measured outwards from the origin of the coordinate system. The displacements $u$, $v$, and $w$ for the layer in the $x$, $\theta$, and $z$ directions respectively are selected as follows:

(1) Linear polynomial in the thickness direction ($z$), using displacement values only at the two nodal surfaces.

(2) Fourier series in the circumferential direction ($\theta$). This is a well-known technique for axisymmetric analysis.

(3) Basic functions and their first derivatives in the axial direction that satisfy the end conditions.

The final displacement functions are written as

$$
\left.\begin{aligned}
u &= \sum_{m=1}^{r} [(1-\bar{z})u_{1m} + \bar{z}u_{2m}] X'_m \cos(n\theta + \theta_0), \\
v &= \sum_{m=1}^{r} [(1-\bar{z})v_{1m} + \bar{z}v_{2m}] X_m \sin(n\theta + \theta_0), \\
w &= \sum_{m=1}^{r} [(1-\bar{z})w_{1m} + \bar{z}w_{2m}] X_m \cos(n\theta + \theta_0),
\end{aligned}\right\} \tag{7.9}
$$

or, written in matrix form,

$$
\{f\} = \begin{Bmatrix} u \\ v \\ w \end{Bmatrix} = \sum [N]_m \{\delta\}_m = [N]\{\delta\} \tag{7.10}
$$

in which $\bar{z} = (z/c)$, $\{\delta\}_m = [u_{1m}v_{1m}w_{1m}u_{2m}v_{2m}w_{2m}]^T$, $n$ is an integer representing the circumferential wave number and $\theta_0$ is phase angle which is introduced for the study of axisymmetric cases (when $n=0$), in which $\theta_0=0$ indicates a motion primarily radial and longitudinal and $\theta_0 = 90°$ indicates a torsional motion.

The strain displacement relationships are given by

$$
\begin{Bmatrix} \varepsilon_x \\ \varepsilon_\theta \\ \varepsilon_z \\ \gamma_{\theta z} \\ \gamma_{zx} \\ \gamma_{x\theta} \end{Bmatrix} = \begin{Bmatrix} \dfrac{\partial u}{\partial x} \\[2mm] \left(1+\dfrac{z}{r_1}\right)^{-1}\left(\dfrac{1}{r_1}\dfrac{\partial v}{\partial \theta}+\dfrac{w}{r_1}\right) \\[2mm] \dfrac{\partial w}{\partial z} \\[2mm] \dfrac{\partial v}{\partial z}+\left(\dfrac{1}{r_1}\dfrac{\partial w}{\partial \theta}-\dfrac{v}{r_1}\right)\Big/\left(1+\dfrac{z}{r_1}\right) \\[2mm] \dfrac{\partial u}{\partial z}+\dfrac{\partial w}{\partial x} \\[2mm] \left(\dfrac{1}{r_1}\dfrac{\partial u}{\partial \theta}\right)\Big/\left(1+\dfrac{z}{r_1}\right)+\dfrac{\partial v}{\partial x} \end{Bmatrix}
$$

$$
= [B]\{\delta\}
$$

$$
= \sum_{m=1}^{r} [B]_m \{\delta\}_m. \tag{7.11}
$$

The stresses corresponding to $\{\varepsilon\}$ are

$$\{\sigma\} = \begin{Bmatrix} \sigma_x \\ \sigma_\theta \\ \sigma_z \\ \tau_{\theta z} \\ \tau_{zx} \\ \tau_{x\theta} \end{Bmatrix} = [D]\{\varepsilon\}$$

$$= [D] \sum_{m=1}^{r} [B]_m \{\delta\}_m \qquad (7.12)$$

with the non-zero elastic constants equal to

$$D_{11} = \eta E_x(1 - \nu_{z\theta}\nu_{\theta z}),$$
$$D_{12} = \eta E_x(\nu_{\theta z} + \nu_{\theta z}\nu_{zx}) = D_{21},$$
$$D_{22} = \eta E_\theta(1 - \nu_{xz}\nu_{zx}),$$
$$D_{23} = \eta E_\theta(\nu_{z\theta} + \nu_{zx}\nu_{x\theta}) = D_{32},$$
$$D_{33} = \eta E_z(1 - \nu_{\theta x}\nu_{x\theta}),$$
$$D_{13} = \eta E_x(\nu_{zx} + \nu_{\theta x}\nu_{z\theta}) = D_{31},$$
$$D_{44} = G_{z\theta}, \quad D_{55} = G_{zx}, \quad D_{66} = G_{x\theta}$$
$$\eta = 1/(1 - \nu_{zx}\nu_{xz} - \nu_{\theta x}\nu_{x\theta} - \nu_{\theta z}\nu_{z\theta}$$
$$- \nu_{\theta z}\nu_{zx}\nu_{x\theta} - \nu_{z\theta}\nu_{\theta x}\nu_{zx})$$

with $\qquad \nu_{ij}E_j = \nu_{ji}E_i \quad (i, j = x, \theta, z).$

Both the stiffness matrix and mass matrix are integrated numerically by Gaussian quadrature.

## (b) Higher order cylindrical layer (HO3)

For such a layer a shape function given by (1.9d) should be used. The displacement given by Nelson et al[4] is of the following form:

$$\left.\begin{aligned} u &= [(1 - 3\bar{r} + 2\bar{r}^2)\,u_1 + (2\bar{r}^2 - \bar{r})\,u_2 + (4\bar{r} - 4\bar{r}^2)u_3]\,\cos(n\theta)\,\sin(\pi z/\lambda), \\ v &= [(1 - 3\bar{r} + 2\bar{r}^2)\,v_1 + (2\bar{r}^2 - \bar{r})\,v_2 + (4\bar{r} - 4\bar{r}^2)v_3]\,\sin(n\theta)\,\cos(\pi z/\lambda), \\ w &= [(1 - 3\bar{r} + 2\bar{r}^2)\,w_1 + (2\bar{r}^2 - \bar{r})\,w_2 + (4\bar{r} - 4\bar{r}^2)w_3]\,\cos(n\theta)\,\cos(\pi z/\lambda), \end{aligned}\right\} (7.13)$$

13

in which $\lambda$ is the axial wave length, and

$$\bar{r} = \frac{r-r_1}{c} = \frac{r-r_1}{r_2-r_1} .$$

For the case of axisymmetric torsion $(n=0)$, it is only necessary to interchange the $\sin(n\theta)$ and $\cos(n\theta)$ terms in the $v$, $u$, and $w$ expressions respectively of (7.13).

The wavelength $\lambda$ is usually not specified, and the natural frequencies are usually given in terms of the axial wave number $\xi = H/\lambda$, where $H$ is the thickness of the cylinder.

The stiffness matrix and mass matrices are given in reference 4.

To demonstrate the accuracy and effectiveness of the present analysis, the numerical results presented here are compared with previous analytical and experimental investigations for free vibration of cylinders over a

TABLE 7.3. DIMENSIONLESS FREQUENCIES $[\Omega = (\omega H/\pi)\sqrt{\varrho/G}]$ FOR FREELY SUPPORTED ISOTROPIC SOLID CYLINDERS $(n = 1,\ H/R = 2.0,$ NO. OF ELEMENTS = 20, POISSON'S RATIO = 0.3)

| $H/L$ | Method | $\Omega_1$ | $\Omega_2$ | $\Omega_3$ | $\Omega_4$ | $\Omega_5$ | $\Omega_6$ |
|-------|--------|-----------|-----------|-----------|-----------|-----------|-----------|
|       | Ref. 6 | 0.000253 | 0.58633 | 0.89677 | 1.69705 | 2.04326 | 2.26038 |
| 0.01  | HO3    | 0.000253 | 0.58627 | 0.89668 | 1.69688 | 2.04305 | 2.26014 |
|       | LO2    | 0.000253 | 0.58643 | 0.89705 | 1.70034 | 2.04821 | 2.26417 |
|       | Ref. 6 | 0.02423  | 0.61082 | 0.90402 | 1.69715 | 2.04855 | 2.26576 |
| 0.1   | HO3    | 0.02424  | 0.61076 | 0.90393 | 1.69698 | 2.04834 | 2.26553 |
|       | LO2    | 0.02423  | 0.61093 | 0.90429 | 1.70043 | 2.05356 | 2.26955 |
|       | Ref. 6 | 0.08703  | 0.67413 | 0.92654 | 1.69829 | 2.06364 | 2.28214 |
| 0.2   | HO3    | 0.08703  | 0.67407 | 0.92645 | 1.69812 | 2.06343 | 2.28190 |
|       | LO2    | 0.08703  | 0.67425 | 0.92681 | 1.70154 | 2.06881 | 2.28589 |
|       | Ref 6  | 0.26658  | 0.83942 | 1.02614 | 1.71296 | 2.11383 | 2.34092 |
| 0.4   | HO3    | 0.26659  | 0.83937 | 1.02606 | 1.71278 | 2.11362 | 2.34070 |
|       | LO2    | 0.26660  | 0.83958 | 1.02642 | 1.71613 | 1.11941 | 2.35085 |
|       | Ref. 6 | 0.46827  | 0.98510 | 1.20161 | 1.76278 | 2.17858 | 2.44666 |
| 0.6   | HO3    | 0.46828  | 0.98503 | 1.20154 | 1.76262 | 2.17837 | 2.44645 |
|       | LO2    | 0.46832  | 0.98529 | 1.20199 | 1.76586 | 2.18440 | 2.45157 |
|       | Ref. 6 | 0.67316  | 1.11984 | 1.39598 | 1.86903 | 2.25398 | 2.56430 |
| 0.8   | HO3    | 0.67317  | 1.11978 | 1.39590 | 1.86890 | 2.25378 | 2.56407 |
|       | LO2    | 0.67327  | 1.12008 | 1.39666 | 1.87195 | 2.25985 | 2.57123 |
|       | Ref. 6 | 0.87660  | 1.26181 | 1.56464 | 2.03841 | 2.34155 | 2.68771 |
| 1.0   | HO3    | 0.87661  | 1.26174 | 1.56456 | 2.03830 | 2.34135 | 2.68750 |
|       | LO2    | 0.87081  | 1.26210 | 1.56577 | 2.04122 | 2.34737 | 2.69715 |

wide range of thickness–radius ratios and with various boundary conditions.

In Table 7.3 it is seen that the frequencies of isotropic freely supported solid cylinders obtained by Armenakas et al.[6] and Nelson et al.[4] using fifty HO3 layers, and, finally, by Cheung and Wu[5] using twenty LO2 layers, are all in excellent agreement This shows that while both types of elements produce accurate results, the higher order finite layer analysis leads to much more complicated algebra and a significantly larger matrix without demonstrating a corresponding improvement in accuracy. Hence the choice of a simple linear variation of displacements through the thickness of a layer is entirely justified.

The LO2 layer was also applied to the analysis of a cylinder with layered materials, and the frequencies obtained (see Table 7.4) are again in

TABLE 7.4. COMPARISON OF NATURAL FREQUENCIES (Hz) FOR A LAYERED FREELY SUPPORTED CYLINDER

| Number of axial half-waves | Number of circumferential waves | Ref. 7 | Ref. 8 | LO2 |
|---|---|---|---|---|
| 1 | 3 | 1148 | 1149 | 1148.5 |
| | 4 | 726 | 726 | 726.2 |
| | 5 | 502 | 502 | 502.2 |
| | 8 | 356 | 356 | 356.4 |
| | 11 | 561 | 562 | 561.9 |
| 2 | 4 | 2146 | 2147 | 2146.3 |
| | 7 | 964 | 965 | 964.9 |
| | 8 | 807 | 807 | 807.2 |
| | 11 | 713 | 714 | 713.8 |
| | 12 | 769 | 770 | 770.2 |
| | 14 | 956 | 956 | 956.8 |
| 3 | 4 | 3398 | 3399 | 3398.4 |
| | 5 | 2719 | 2720 | 2718.9 |
| | 7 | 1793 | 1794 | 1793.7 |

Data:

| Layer | Thickness (m) | Number of elements | $E$ (kN/m$^2$) | Poisson's ratio | Density (kN/s$^2$/m$^4$) |
|---|---|---|---|---|---|
| Inside | 0.000254 | 10 | $4.8265 \times 10^6$ | 0.30 | 1.31425 |
| Middle | 0.000127 | 5 | $206.85 \times 10^6$ | 0.30 | 8.52666 |
| Outside | 0.000254 | 10 | $4.8265 \times 10^6$ | 0.30 | 1.31425 |

$L = 0.2184$ m; $R = 0.1016$ m

TABLE 7.5. COMPARISON OF NATURAL FREQUENCIES (Hz) BETWEEN EXPERIMENTAL RESULTS AND PRESENT ANALYSIS

| Number of circumferential waves | Number of axial half waves | Freely supported[a] | | Clamped-clamped[a] | | Clamped-free[b] | | Free-free[c] | |
|---|---|---|---|---|---|---|---|---|---|
| | | Ref. 9 | LO2 | Ref. 9 | LO2 | Ref. 9 | LO2 | Ref. 9 | LO2 |
| 2 | 1 | | 824.5 | | 1071.0 | | 329.2 | 7.2 | 7.2 |
| | 2 | | 1892.9 | | 2005.0 | | 1217.2 | 9.2 | 8.7 |
| | 3 | | 2509.4 | | 2563.7 | | 2188.1 | | 1471.0 |
| 3 | 1 | | 461.5 | | 708.8 | 155.0 | 174.8 | 20.1 | 20.5 |
| | 2 | | 1292.8 | | 1462.2 | | 770.1 | 22.3 | 20.8 |
| | 3 | | 1961.6 | | 2038.3 | | 1578.1 | | 906.4 |
| 4 | 1 | 287.0 | 287.6 | | 496.5 | 107.0 | 111.8 | 37.9 | 39.4 |
| | 2 | | 900.9 | | 1092.7 | | 519.0 | 40.3 | 42.0 |
| | 3 | | 1511.9 | | 1609.4 | | 1149.0 | | 576.1 |
| 5 | 1 | 203.0 | 201.8 | 321.0 | 366.0 | 89.0 | 94.6 | 61.9 | 63.7 |
| | 2 | | 651.6 | | 838.6 | 341.0 | 372.5 | 65.0 | 66.5 |
| | 3 | | 1172.1 | | 1284.5 | | 860.2 | 367.0 | 395.7 |
| 6 | 1 | 175.0 | 166.6 | 263.0 | 289.1 | 102.0 | 106.3 | 91.6 | 93.4 |
| | 2 | | 493.6 | | 662.2 | 276.0 | 288.5 | 94.6 | 96.3 |
| | 3 | | 924.3 | | 1043.3 | | 666.3 | 292.0 | 297.6 |
| 7 | 1 | 163.0 | 166.2 | 233.0 | 249.9 | 130.0 | 134.7 | 125.0 | 128.6 |
| | 2 | | 396.7 | 479.0 | 541.3 | 240.0 | 248.1 | 129.0 | 131.6 |
| | 3 | | 747.4 | | 865.4 | | 537.4 | 242.0 | 251.8 |
| 8 | 1 | 188.0 | 189.3 | 227.0 | 242.1 | 166.0 | 172.6 | 165.0 | 169.2 |
| | 2 | 345.0 | 345.0 | 418.0 | 463.1 | 227.0 | 241.5 | 170.0 | 172.2 |
| | 3 | | 625.2 | | 736.4 | | 456.8 | 235.0 | 243.6 |

| | | | | | | | | | |
|---|---|---|---|---|---|---|---|---|---|
| 9 | 1 | 224.0 | 226.9 | 244.0 | 259.2 | 208.0 | 217.4 | 210.0 | 215.2 |
| | 2 | | 329.2 | | 420.7 | 246.0 | 260.3 | 214.0 | 218.2 |
| | 3 | | 547.1 | | 647.2 | 400.0 | 415.0 | 252.0 | 262.5 |
| 10 | 1 | 268.0 | 274.1 | 281.0 | 294.3 | 260.0 | 268.3 | 262.0 | 266.6 |
| | 2 | 339.0 | 341.7 | | 409.1 | 281.0 | 296.7 | 266.0 | 269.7 |
| | 3 | | 506.4 | | 592.4 | | 405.5 | 290.0 | 299.4 |
| 11 | 1 | 326.0 | 328.6 | 283.0 | 342.0 | 317.0 | 324.8 | 318.0 | 323.4 |
| | 2 | 369.0 | 374.9 | 397.0 | 423.2 | 337.0 | 345.3 | 322.0 | 326.5 |
| | 3 | 491.0 | 497.4 | 562.0 | 568.0 | 409.0 | 422.2 | 339.0 | 348.4 |

Data: five identical layers; no. of terms of $O(x)$ series = 5;

$E = 6.895 \times 10^7$ kN/m$^2$; Poisson's ratio = 0.30; density = 2.714 kN s$^2$/m$^4$

$R = 0.24227$ m; $H = 0.0006477$ m; $L = 0.6096$ m[a] (0.6255 m[b], 0.6382 m[c])

very good agreement with those presented by Weingarten[7] and by Bert et al.[8]

For cylinders with various boundary conditions, the results obtained from LO2 finite layer analysis are compared with experimental measurements in Table 7.5. Reasonable agreements can be observed. Only thin shells have been analysed in this example because no previous investigations on free vibrations of thick cylinders having any conditions other than simply supported are available for comparison.

## 7.3. FINITE PRISM METHOD

### 7.3.1. INTRODUCTION

In the finite element approach it is often convenient to define the coordinates of the more complex element shapes by the nodal coordinates $\xi_k$,

$$\xi = \sum_{k=1}^{s} \phi_k \xi_k \qquad (7.14a)$$

in which $s$ refers to the number of nodes of the element.

Similarly, the displacements $\delta$ of a point within the element can also be expressed in terms of the nodal displacements $\delta_k$,

$$\delta = \sum_{k=1}^{s} \psi_k \delta_k, \qquad (7.14b)$$

where $\phi_k$ and $\psi_k$ are functions of a coordinate system (Cartesian, skew, curvilinear, etc.). These two functions are, in general, not the same. For the special case in which $\phi_k$ and $\psi_k$ are identical, the element is termed isoparametric.

The concept of isoparametric elements was first introduced by Taig[10] and by Irons[11] and later extended by Ergatoudis et al.[12-14] In this section the concept is further extended to the formulation of straight and curved simply supported finite prisms[15] with eight-node isoparametric sections.

## 7.3.2. STRAIGHT AND CURVED FINITE PRISMS
### WITH EIGHT-NODE ISOPARAMETRIC SECTIONS

Referring to Fig. 7.2 and eqn. (1.1b) it can be seen that in the plane of the cross-section the shape functions given in (1.9g) should be used, while in the axial direction $Y_m$ there are again some Fourier series which

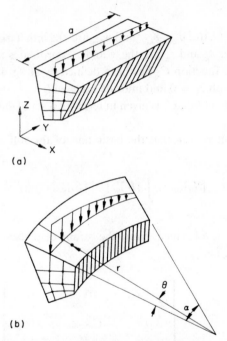

(a)

(b)

FIG. 7.2. Typical prismatic situations. (a) Straight prism. (b) Curved prism.

satisfy the simple support end conditions $u = w = \partial v/\partial y = 0$. A suitable set of displacement functions for a straight prism is

$$
\left.
\begin{aligned}
u &= \sum_{m=1}^{r} \sum_{k=1}^{8} C_k u_{km} \sin k_m y, \\
v &= \sum_{m=1}^{r} \sum_{k=1}^{8} C_k v_{km} \cos k_m y, \\
w &= \sum_{m=1}^{r} \sum_{k=1}^{8} C_k w_{km} \sin k_m y.
\end{aligned}
\right\}
\tag{7.15}
$$

The coordinates of the eight-node isoparametric section can thus be similarly defined as

$$x = \sum_{k=1}^{8} C_k x_k, \\ z = \sum_{k=1}^{8} C_k z_k. \qquad (7.16)$$

In (7.15) and (7.16), $u_{km}$, $v_{km}$, and $w_{km}$ are the $m$th term displacements at the $k$th node, $x_k$ and $z_k$ are the $x$ and $z$ coordinates at the $k$th node, while the shape function $C_k$ has been defined by (1.9g) for corner node, midside node with $\xi_k = 0$ and midside node with $\eta_k = 0$ respectively. As can be seen from (1.9g), $C_k$ is given in terms of a set of curvilinear coordinates $\xi$ and $\eta$.

From (1.32) it is seen that the basic unit of the stiffness matrix is

$$[S_{ij}]_{mn} = \int [B_i]_m^T [D] [B_j]_n \, d \, (\text{vol.})$$

in which $[B_i]_m$ is obtained from the strain displacement relationship given by (7.3a), i.e.

$$[B_i]_m = \begin{bmatrix} \dfrac{\partial C_i}{\partial x} s_m & 0 & 0 \\[2ex] 0 & -C_i k_m s_m & 0 \\[2ex] 0 & 0 & \dfrac{\partial C_i}{\partial z} s_m \\[2ex] C_i k_m c_m & \dfrac{\partial C_i}{\partial x} c_m & 0 \\[2ex] 0 & \dfrac{\partial C_i}{\partial z} c_m & C_i k_m c_m \\[2ex] \dfrac{\partial C_i}{\partial x} s_m & 0 & \dfrac{\partial C_i}{\partial z} s_m \end{bmatrix} \qquad (7.17)$$

where $s_m = \sin k_m y$ and $c_m = \cos k_m y$.

The elasticity matrix for orthotropic materials is

$$[D] = \begin{bmatrix} D_{11} & D_{12} & D_{13} & 0 & 0 & 0 \\ D_{21} & D_{22} & D_{23} & 0 & 0 & 0 \\ D_{31} & D_{32} & D_{33} & 0 & 0 & 0 \\ 0 & 0 & 0 & D_{44} & 0 & 0 \\ 0 & 0 & 0 & 0 & D_{55} & 0 \\ 0 & 0 & 0 & 0 & 0 & D_{66} \end{bmatrix} \qquad (7.18)$$

If $[B_i]_m$ and $[D]$ are now substituted into (1.32), then we have

$$[S_{ij}]_{mn} = \int_A \int_0^a$$

$$\begin{bmatrix} D_{11} \dfrac{\partial C_i}{\partial x} \dfrac{\partial C_j}{\partial x} s_m s_n & -D_{12} C_j \dfrac{\partial C_i}{\partial x} k_m s_m s_n & D_{13} \dfrac{\partial C_i}{\partial x} \dfrac{\partial C_j}{\partial z} s_m s_n \\ +D_{44} C_i C_j k_m k_n c_m c_n & +D_{44} C_i \dfrac{\partial C_j}{\partial x} k_m s_m s_n & +D_{66} \dfrac{\partial C_i}{\partial x} \dfrac{\partial C_j}{\partial z} s_m s_n \\ +D_{66} \dfrac{\partial C_i}{\partial x} \dfrac{\partial C_j}{\partial x} s_m s_n & & \\[2ex] -D_{21} C_i \dfrac{\partial C_j}{\partial x} k_m s_m s_n & D_{22} C_i C_j k_m k_n s_m s_n & -D_{23} C_i \dfrac{\partial C_j}{\partial z} k_m s_m s_n \\ +D_{44} \dfrac{\partial C_i}{\partial x} C_j k_n c_m c_n & +D_{44} \dfrac{\partial C_i}{\partial x} \dfrac{\partial C_j}{\partial x} c_m c_n & +D_{55} \dfrac{\partial C_i}{\partial z} C_j k_n s_m s_n \\ & +D_{55} \dfrac{\partial C_i}{\partial z} \dfrac{\partial C_j}{\partial z} c_m c_n & \\[2ex] D_{31} \dfrac{\partial C_i}{\partial z} \dfrac{\partial C_j}{\partial x} s_m s_n & -D_{32} \dfrac{\partial C_i}{\partial z} C_j k_n s_m s_n & D_{33} \dfrac{\partial C_i}{\partial z} \dfrac{\partial C_j}{\partial z} s_m s_n \\ +D_{66} \dfrac{\partial C_i}{\partial z} \dfrac{\partial C_j}{\partial x} s_m s_n & +D_{55} C_i \dfrac{\partial C_j}{\partial z} k_m c_m c_n & +D_{55} C_i C_j k_m k_n C_m c_n \\ & & +D_{66} \dfrac{\partial C_i}{\partial z} \dfrac{\partial C_j}{\partial z} s_m s_n \end{bmatrix}$$

$$dx\, dz\, dy \quad (7.19)$$

All the coefficients inside the submatrix $[S_{ij}]_{mn}$ contain integrals of the form $\int_0^a \sin k_m y \sin k_n y \, dy$ or $\int_0^a \cos k_m y \cos k_n y \, dy$ and will vanish for $m \neq n$, consequently the different terms of the series are uncoupled.

It is interesting to note the decoupling will not occur for general anisotropic materials with twenty-one constants[16] because of the presence of the integral $\int_0^a \sin k_m y \cos k_m y \, dy$ which does not exhibit orthogonal properties.

A closer examination of (7.16) shows that while the shape function $C_i$ is given in terms of local curvilinear coordinates $\xi$ and $\eta$, all differentiation and integration are in terms of the global Cartesian coordinate system since all the strains are written as derivatives of $x$, $y$, and $z$. Hence it is necessary to (i) rewrite the derivatives with respect to local coordinate system, i.e. transform $\partial C_i/\partial x$, $\partial C_i/\partial z$ to $\partial C_i/\partial \xi$, $\partial C_i/\partial \eta$, (ii) change the area $dxdz$ to $d\xi d\eta$, and (iii) change the integration limits to suit the local coordinate system.

By applying the chain rule in partial differentiation it is possible to write

$$
\left.
\begin{aligned}
\frac{\partial C_i}{\partial \xi} &= \frac{\partial C_i}{\partial x}\frac{\partial x}{\partial \xi} + \frac{\partial C_i}{\partial z}\frac{\partial z}{\partial \xi}, \\
\frac{\partial C_i}{\partial \eta} &= \frac{\partial C_i}{\partial x}\frac{\partial x}{\partial \eta} + \frac{\partial C_i}{\partial z}\frac{\partial z}{\partial \eta}.
\end{aligned}
\right\}
\tag{7.20}
$$

Thus writing in matrix form we have

$$
\left\{
\begin{array}{c}
\dfrac{\partial C_i}{\partial \xi} \\[2mm]
\dfrac{\partial C_i}{\partial \eta}
\end{array}
\right\}
=
\begin{bmatrix}
\dfrac{\partial x}{\partial \xi} & \dfrac{\partial z}{\partial \xi} \\[2mm]
\dfrac{\partial x}{\partial \eta} & \dfrac{\partial z}{\partial \eta}
\end{bmatrix}
\left\{
\begin{array}{c}
\dfrac{\partial C_i}{\partial x} \\[2mm]
\dfrac{\partial C_i}{\partial \eta}
\end{array}
\right\}
= [J]
\left\{
\begin{array}{c}
\dfrac{\partial C_i}{\partial x} \\[2mm]
\dfrac{\partial C_i}{\partial z}
\end{array}
\right\}.
\tag{7.21}
$$

The matrix $[J]$, referred to as the Jacobian matrix, can be written explicitly for isoparametric elements by virtue of (7.16). The expanded form of the Jacobian matrix is

$$[J] = \begin{bmatrix} \dfrac{\partial x}{\partial \xi} & \dfrac{\partial z}{\partial \xi} \\[2mm] \dfrac{\partial x}{\partial \eta} & \dfrac{\partial z}{\partial \eta} \end{bmatrix} = \sum_{k=1}^{8} \begin{bmatrix} \dfrac{\partial C_k}{\partial \xi} x_k & \dfrac{\partial C_k}{\partial \xi} z_k \\[2mm] \dfrac{\partial C_k}{\partial \eta} x_k & \dfrac{\partial C_k}{\partial \eta} z_k \end{bmatrix}$$

$$= \begin{bmatrix} \dfrac{\partial C_1}{\partial \xi} & \dfrac{\partial C_2}{\partial \xi} & \dfrac{\partial C_3}{\partial \xi} & \dfrac{\partial C_4}{\partial \xi} & \dfrac{\partial C_5}{\partial \xi} & \dfrac{\partial C_6}{\partial \xi} & \dfrac{\partial C_7}{\partial \xi} & \dfrac{\partial C_8}{\partial \xi} \\[2mm] \dfrac{\partial C_1}{\partial \eta} & \dfrac{\partial C_2}{\partial \eta} & \dfrac{\partial C_3}{\partial \eta} & \dfrac{\partial C_4}{\partial \eta} & \dfrac{\partial C_5}{\partial \eta} & \dfrac{\partial C_6}{\partial \eta} & \dfrac{\partial C_7}{\partial \eta} & \dfrac{\partial C_8}{\partial \eta} \end{bmatrix} \begin{bmatrix} x_1 & z_1 \\ x_2 & z_2 \\ x_3 & z_3 \\ x_4 & z_4 \\ x_5 & z_5 \\ x_6 & z_6 \\ x_7 & z_7 \\ x_8 & z_8 \end{bmatrix}$$

$$(7.22)$$

With $[J]$ computed it is possible to write the global derivatives as

$$\begin{Bmatrix} \dfrac{\partial C_i}{\partial x} \\[2mm] \dfrac{\partial C_i}{\partial z} \end{Bmatrix} = [J]^{-1} \begin{Bmatrix} \dfrac{\partial C_i}{\partial \xi} \\[2mm] \dfrac{\partial C_i}{\partial \eta} \end{Bmatrix}. \qquad (7.23)$$

The elemental area $dx\,dz$, if expressed in the curvilinear coordinate system, becomes

$$dx\,dz = \det |J|\, d\xi\, d\eta. \qquad (7.24)$$

Finally, the integration limits are changed to $+1$ and $-1$ since $\xi$ and $\eta$ take up such values at the four sides of the quadrilateral section. This particular property is quite valuable because Gaussian quadrature can be applied directly.

The extension to circular prisms is obvious, and the displacement functions given in (7.13) requires only a small modification which consists of replacing the variable $y$ by the angle $\theta$ and the span $a$ by the subtended angle $\alpha$. Thus

$$
\left.
\begin{aligned}
u &= \sum_{m=1}^{r} \sum_{k=1}^{8} C_k u_{km} \sin \frac{m\pi\theta}{\alpha}, \\
v &= \sum_{m=1}^{r} \sum_{k=1}^{8} C_k v_{km} \cos \frac{m\pi\theta}{\alpha}, \\
w &= \sum_{m=1}^{r} \sum_{k=1}^{8} C_k w_{km} \sin \frac{m\pi\theta}{\alpha},
\end{aligned}
\right\}
\tag{7.25}
$$

with $C_k$ being given by (1.9g) as before.

The strain displacement relationship has been given in (7.11). The formulation of the stiffness matrix can now proceed along the same lines as for the straight prism.

In order to assess the accuracy of the finite prism method, a simply supported straight beam subject to a central concentrated load (Fig. 7.3) is investigated using different mesh size and eleven odd harmonics. The displacements and stresses for meshes (a), (b), (c) of Fig. 7.3 are listed in Table 7.6 and compared with an exact solution.[17] The comparison

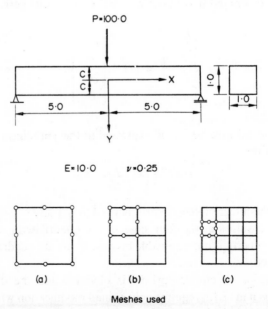

FIG. 7.3. Simply supported beam under point load (Too[15]).

TABLE 7.6. COMPARISON OF STRESSES OF A SIMPLY SUPPORTED BEAM UNDER POINT LOAD (Too[15])

| Stress | $y$ | Mesh | | | Exact |
|---|---|---|---|---|---|
| | | (a) | (b) | (c) | |
| $\sigma_x$ at midspan | $-c$ | $-1571.03$ | $-1907.69$ | $-2288.86$ | $\infty$ |
| | $-c/2$ | — | $-770.75$ | $-749.02$ | $-664.4$ |
| | 0 | 71.69 | 3.63 | 18.89 | 24.2 |
| | $c/2$ | — | 730.31 | 727.11 | 722.8 |
| | $c$ | 1395.45 | 1465.30 | 1475.55 | 1473.4 |
| $\sigma_y$ at midspan | $-c$ | $-296.56$ | $-985.72$ | $-1847.60$ | $\infty$ |
| | $-c/2$ | — | $-558.94$ | $-579.23$ | $-246.0$ |
| | 0 | 56.63 | $-184.15$ | $-103.04$ | $-91.2$ |
| | $c/2$ | — | $-23.45$ | $-29.60$ | $-29.0$ |
| | $c$ | $-83.42$ | 43.16 | 9.80 | 00.0 |
| $\tau_{xy}$ at $\frac{1}{4}$ span | $-c$ | 40.81 | 10.01 | 1.20 | 0.00 |
| | $-c/2$ | — | 47.62 | 55.24 | 56.25 |
| | 0 | 60.98 | 80.76 | 73.40 | 75.00 |
| | $c/2$ | — | 47.67 | 55.10 | 56.20 |
| | $c$ | 40.88 | 10.98 | 2.63 | 0.00 |
| $w$ at midspan | | 2545.15 | 2562.42 | 2563.90 | 2567.28 |

can be regarded as satisfactory for the coarse mesh (a) and good for the fine mesh (c).

It is interesting to note that local effects can be predicted with reasonable accuracy for a small cost.

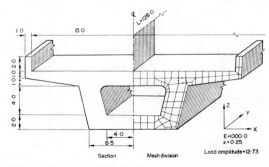

FIG. 7.4. Dimension and meshes of a simply supported thick bridge box (Too[15]).

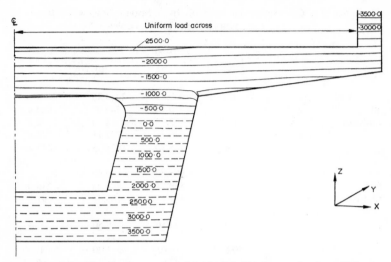

FIG. 7.5A. Simply supported straight thick bridge box under sinusoidal load. $Y$-stress at mid-span, $(\sigma_y) L/2$.

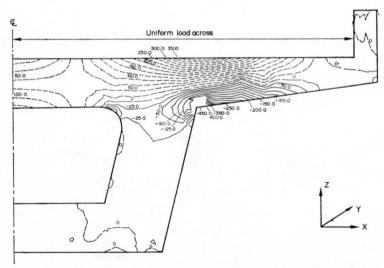

FIG. 7.5B. Simply supported straight thick bridge box under sinusiodal load. $X$-stress at mid-span, $(\sigma_x) L/2$.

The second example is of a more practical nature. A thick-walled straight box girder bridge was analysed under a sinusoidal load which was uniformly distributed in the transverse direction. The dimensions of the bridge and the mesh used are shown in Fig. 7.4. Since the load distribution is proportional to $\sin(\pi y/a)$, in the longitudinal direction, only the stiffness matrix corresponding to the first term of the series needs to be solved. The stresses obtained are presented as stress contours in Figs. 7.5A, B, and C. Fig. 7.5A shows the $\sigma_y$ stress normal to the plane of the section, and it can be seen that the assumption of linear stress distribution in simple beam theory holds quite well here. In Fig. 7.5B is

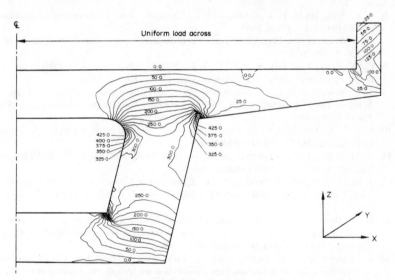

FIG. 7.5C. Simply supported straight thick bridge box under sinusoidal load. $YZ$-stress at quarter-span, $(\tau_{yz})L/4$.

shown the transverse stress $\sigma_x$, and here it is possible to see the transverse bending of the upper flange and also the high stress concentration at the bottom root of the cantilever slab. In Fig. 7.5C it is of interest to observe that the major part of the transverse shear stress $\tau_{yz}$ are carried by the web panels and edge beams, as was expected.

## REFERENCES

1. Y. K. CHEUNG and S. CHAKRABARTI, Analysis of simply supported thick, layered plates, *Am. Soc. Civil Engrs* **97**, EM3, 1039–44 (June 1971).
2. Y. K. CHEUNG and S. CHAKRABARTI, Free vibration of thick, layered rectangular plates by a finite layer method, *J. Sound Vibration* **21** (3) 277–84 (1972).
3. G. B. WARBURTON, The vibration of rectangular plates, *Proceedings of the Institution of Mechanical Engineers* **168** (12) 371–84 (1954).
4. R. B. NELSON, S. B. DONG, and R. D. KALRA, Vibrations and waves in laminated orthotropic circular cylinders, *J. Sound Vibration* **18** (3) 429–44 (1971).
5. Y. K. CHEUNG and C. I. WU, Free vibrations of thick, layered cylinders having finite length with various boundary conditions, *J. Sound Vibration* **24** (2) 189–200 (1972).
6. A. E. ARMENAKAS, D. C. GAZIS, and G. HERRMANN, *Free Vibrations of Cylindrical Shells*, Pergamon Press, 1969.
7. V. I. WEINGARTEN, Free vibrations of multilayered cylindrical shells, *Exp. Mech.* **4**, 200–5 (1964).
8. C. W. BERT, J. L. BAKER, and D. M. EGLE, Free vibrations of multilayer anisotropic cylindrical shells, *J. Composite Materials* **3**, 480–99 (1969).
9. J. L. SEWALL and E. C. NAUMANN, *An Experimental and Analytical Vibration Study of Thin Cylindrical Shells With and Without Longitudinal Stiffeners*, NASA TN D-4705, 1968.
10. I. C. TAIG, *Structure Analysis by Displacement Method*, English Electric Aviation Ltd., Report SO 17 (1961) unpublished.
11. B. M. IRONS, *Stress Analysis of Stiffnesses using Numerical Integration*, Rolls Royce Ltd., ASM 622 June 1963 (internal report).
12. J. ERGATOUDIS, B. M. IRONS and O. C. ZIENKIEWICZ, Curved isoparametric quadrilateral elements for finite element analysis, *Int. J. Solids Structures* **7**, 31–42 (1968).
13. B. M. IRONS and O. C. ZIENKIEWICZ, The isoparametric system—a new concept in finite element analysis, *Conference—Recent Advances in Stress Analysis*, JBCSA, Royal Aeronautical Society London, March 1968.
14. J. ERGATOUDIS, B. M. IRONS, and O. C. ZIENKIEWICZ, Three dimensional analysis of arch dams and their foundations, *Symposium on Arch Dams, Institution of Civil Engineers, London, March 1968*.
15. J. J. M. TOO, Two-Dimension, Plate, Shell and Finite Prism Isoparametric Elements and Their Applications, PhD thesis, Department of Civil Engineering, University of Wales, Swansea, 1971.
16. R. L. BISPLINGHOFF, J. W. MAR, and T. H. H. PIAN, *Statics of Deformable Solids*, Addison-Wesley, 1965.
17. S. TIMOSHENKO and GOODIER, *Theory of Elasticity*, 2nd edn., McGraw-Hill, 1951.

# CHAPTER 8

# Computation methods and computer program

## 8.1. INTRODUCTION

In this chapter the various computation methods used in the solution of a structural problem by the finite strip method will be discussed in detail. A computer program for the analysis of folded plate structures will also be presented together with some of the intermediate and final output of an example. It should be mentioned that the techniques presented here are quite general and can be applied to the analysis of other problems such as plates, shells, solids, etc.

The finite strip analysis of a structure can be divided into a number of steps as follows:

(i)     Data preparation.
(ii)    Computation of individual stiffness matrix, transformation matrix, and global stiffness matrix.
(iii)   Assemblage of stiffness matrices.
(iv)    Introduction of prescribed displacement conditions.
(v)     Solution of simultaneous equations.
(vi)    Calculation of internal forces.

Steps (i) – (iv) are common for both static and dynamic problems. For frequency analysis steps (v) and (vi) should be replaced by a single step of eigenvalue solution which will be discussed separately.

The various steps listed above are to be explained in conjunction with the analysis of a folded plate structure shown in Fig. 8.1. For brevity, only the first harmonic will be used.

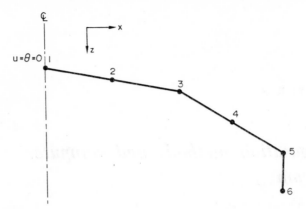

FIG. 8.1. Division into finite strips and boundary conditions:

## 8.2. DATA PREPARATION

In this section a sketch diagram of the middle lines of the cross-section is drawn with both the global axes of the structure and individual axes for each strip indicated. It is customary to use an arrow to indicate the direction of each individual $x'$ axis.

Based on the sketch, the following data can be tabulated and later fed into an electronic computer as input.

(1) Node numbers and the corresponding $x$, $z$ coordinates. Note that the difference of the node numbers between two nodes which are connected with each other determines the bandwidth of the overall matrix and should therefore be kept as small as possible in the numbering scheme.

(2) Element numbers and the corresponding node numbers of each strip. The elastic and geometric properties and the distributed load of each strip should also be listed.

(3) Line loads or point loads at the nodal lines.

(4) Prescribed displacements of certain nodes.

The input data for the folded plate structure of Fig. 4.4 with the mesh shown in Fig. 8.1 are tabulated as follows:

*Nodal coordinates*

| Node | $x$ (m) | $z$ (m) |
|------|---------|---------|
| 1 | 0.000 | 0.000 |
| 2 | 4.915 | 0.875 |
| 3 | 9.830 | 1.750 |
| 4 | 14.165 | 4.250 |
| 5 | 18.500 | 6.750 |
| 6 | 18.500 | 9.750 |

*Strip properties*

| Strip | Left-hand node | Right-hand node | Thickness (m) | Distributed load (N/m²) |
|-------|-----------|------------|-----------|-------------|
| 1 | 1 | 2 | 0.25 | 80 |
| 2 | 2 | 3 | 0.25 | 80 |
| 3 | 3 | 4 | 0.25 | 80 |
| 4 | 4 | 5 | 0.25 | 80 |
| 5 | 5 | 6 | 0.50 | 75 |

*Prescribed displacements*

| Node | $u$ | $v$ | $w$ | $\theta$ |
|------|-----|-----|-----|----------|
| 1 | 0 | | | 0 |

*Material properties and length of strip*

| $E_x$ | $E_y$ | $\nu_x$ | $\nu_y$ | $G$ | Length (m) |
|-------|-------|---------|---------|-----|------------|
| 1 | 1 | 0 | 0 | 0.5 | 70 |

## 8.3. INDIVIDUAL STIFFNESS MATRIX AND TRANSFORMATION MATRIX

The individual stiffness matrix of a strip is normally computed with respect to the local coordinate system and then transformed to the global coordinate system (if necessary). An inspection of a stiffness matrix will show that all the coefficients are functions of the material and geometric properties of a strip. The various items, with the exception of the widths of the strips, have all been tabulated in Section 8.2. However, it is possible to compute the width from the coordinates of the two nodes associated with a strip. As an example, the width of strip (1) in Fig. 8.1 is

$$
\begin{aligned}
b_1 &= \sqrt{(x(2)-x(1))^2+(z(2)-z(1))^2} \\
&= \sqrt{(4.915-0)^2+(0.875-0)^2} \\
&= 4.992 \text{ m.}
\end{aligned} \tag{8.1}
$$

The transformation matrix is given by (4.8) and (4.9), with the direction cosines also computed from the nodal coordinates. Thus for strip (1)

$$
\cos \beta_1 = \frac{x(2)-x(1)}{b_1} = \frac{4.915}{4.992} = 0.9845
$$

and

$$
\sin \beta_1 = \frac{z(2)-z(1)}{b_1} = \frac{0.875}{4.992} = 0.1753. \tag{8.2}
$$

In order to save computer storage, the transformation procedure described in Section 4.3 is usually modified in the following manner:

$$
\begin{aligned}
\underset{(8\times8)}{[S]} &= \underset{(8\times8)}{[R]^T} \quad \underset{(8\times8)}{[S']} \quad \underset{(8\times8)}{[R]} \\[2mm]
&= \begin{bmatrix} [r]^T & [0] \\ [0] & [r]^T \end{bmatrix} \begin{bmatrix} [S'_{11}] & [S'_{12}] \\ [S'_{21}] & [S'_{22}] \end{bmatrix} \begin{bmatrix} [r] & [0] \\ [0] & [r] \end{bmatrix} \\[2mm]
&= \begin{bmatrix} [r]^T \ [S'_{11}] \ [r] & [r]^T \ [S'_{12}] \ [r] \\ [r]^T \ [S'_{21}] \ [r] & [r]^T \ [S'_{22}] \ [r] \end{bmatrix}
\end{aligned} \tag{8.3}
$$

or

$$
\underset{4(\times4)}{[S_{ii}]} = \underset{(4\times4)}{[r]^T} \quad \underset{(4\times4)}{[S_{ij}]} \quad \underset{(4\times4)}{[r]} \tag{8.4}
$$

For the folded plate example, the transformation matrix of strip (1) is

$$
[r]_{(1)} = \begin{bmatrix} 0.9845 & 0 & -0.1753 & 0 \\ 0 & 1 & 0 & 0 \\ 0.1753 & 0 & 0.9845 & 0 \\ 0 & 0 & 0 & 1 \end{bmatrix} \tag{8.5}
$$

and the corresponding individual stiffness matrix [see (4.2)] in local coordinates is

$$[S'_{11}]_{(1)} = \begin{bmatrix} 1.7674 & -0.0982 & 0 & 0 \\ -0.0982 & 0.9057 & 0 & 0 \\ 0 & 0 & 0.0044 & 0.0110 \\ 0 & 0 & 0.0110 & 0.0366 \end{bmatrix} \quad (8.6)$$

$$[S'_{21}]_{(1)} = \begin{bmatrix} -1.7454 & -0.0982 & 0 & 0 \\ 0.0982 & -0.8617 & 0 & 0 \\ 0 & 0 & -0.0044 & -0.0110 \\ 0 & 0 & 0.0110 & 0.0182 \end{bmatrix} = [S'_{12}]^T_{(1)} \quad (8.7)$$

$$[S'_{22}]_{(1)} = \begin{bmatrix} 1.7674 & 0.982 & 0 & 0 \\ 0.0982 & 0.9057 & 0 & 0 \\ 0 & 0 & 0.0044 & -0.0110 \\ 0 & 0 & -0.0110 & 0.0366 \end{bmatrix} \quad (8.8)$$

After transformation, the stiffness matrix becomes

$$[S_{11}]_{(1)} = \begin{bmatrix} 1.7132 & -0.0967 & 0.3042 & -0.0019 \\ -0.0967 & 0.9057 & -0.0172 & 0 \\ 0.3042 & -0.0172 & 0.0586 & 0.0108 \\ -0.0019 & 0 & 0.0108 & 0.0366 \end{bmatrix} \quad (8.9)$$

$$[S_{21}]_{(1)} = \begin{bmatrix} -1.6919 & -0.0967 & -0.3004 & 0.0019 \\ 0.0967 & -0.8617 & 0.0172 & 0 \\ -0.3004 & -0.0172 & -0.0579 & -0.0108 \\ -0.0019 & 0 & 0.0108 & 0.0182 \end{bmatrix} \quad (8.10)$$

$$[S_{22}]_{(1)} = \begin{bmatrix} 1.7132 & 0.0967 & 0.3042 & 0.0019 \\ 0.0967 & 0.9057 & 0.0172 & 0 \\ 0.3042 & 0.0172 & 0.0586 & -0.0108 \\ 0.0019 & 0 & -0.0108 & 0.0366 \end{bmatrix} \quad (8.11)$$

The stiffness matrices for all the other strips can be worked out in a similar manner.

### .4. ASSEMBLAGE OF STIFFNESS MATRICES

The transformed stiffness matrices of all the strips are now combined or assembled to form the final stiffness matrix of the structure. This assembly is carried out through inspection.

Because of the existence of the eigenfunction series in the displacement function and the coupling between the terms of the series, the finite strip stiffness matrix is very often somewhat more complicated than that of a standard finite element, and there are three different situations which will be discussed separately in the subsequent paragraphs.

### (a) Finite strip with simply supported ends

For simply supported strips, the terms of the series are uncoupled and the stiffness matrices of each term can be formed, assembled, and solved separately. Thus if nodes 1 and 2 of strip (i) is associated with nodes $I$ and $J$ of the structure respectively, then for the $m$th term of the series, the four submatrices of the stiffness matrix $S_{mm\,(i)}$ will be added into the framework of the overall stiffness matrix as follows:

| Nodes $I-1$ $I-1$ | $I$ | $I+1...$ $...J-1$ | $J$ | $J-1$ |
|---|---|---|---|---|
| $I$ | $[S_{11}]_{mm\,(i)}$ | | $[S_{12}]_{mm\,(i)}$ | |
| $I+1$ $J-1$ | | | | |
| $J$ | $[S_{21}]_{mm\,(i)}$ | | $[S_{22}]_{mm\,(i)}$ | |
| $J+1$ | | | | |

$$(8.12)$$

The folded plate example belongs to this category, and the overall stiffness matrix (for the $m$th term) is of the form given in (8.13).

### (b) Finite strip with general support conditions (8.14)

For finite strip with end conditions other than simply supported, the various terms of the series will no longer decouple, and the stiffness matrix now takes up the general form given by (1.29). The assembly of the stiffness matrix of the same strip (i) will now be demonstrated for a typical submatrix $[S]_{mn\,(i)}$ [(1.30)]. The coefficients $m$ and $n$ should vary from 1 to $r$, in which $r$ is the total number of terms used in the analysis.

### (c) Finite layer with general support conditions (8.16)

The finite layer formulation, which is described in Chapter 7, uses displacement functions consisting of a polynomial in one direction and series in the other two orthozonal directions. The general form of the stiffness matrix is given by (7.6).

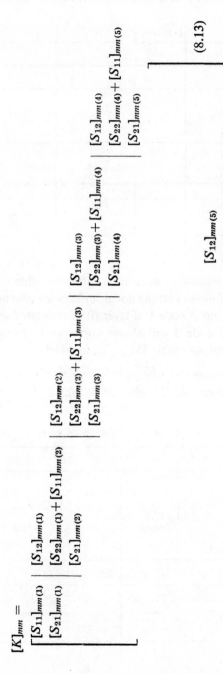

$$[K]_{mm} =
\begin{bmatrix}
[S_{11}]_{mm\,(1)} & [S_{12}]_{mm\,(1)} \\
[S_{21}]_{mm\,(1)} & [S_{22}]_{mm\,(1)} + [S_{11}]_{mm\,(2)} & [S_{12}]_{mm\,(2)} \\
& [S_{21}]_{mm\,(2)} & [S_{22}]_{mm\,(2)} + [S_{11}]_{mm\,(3)} & [S_{12}]_{mm\,(3)} \\
& & [S_{21}]_{mm\,(3)} & [S_{22}]_{mm\,(3)} + [S_{11}]_{mm\,(4)} & [S_{12}]_{mm\,(4)} \\
& & & [S_{21}]_{mm\,(4)} & [S_{22}]_{mm\,(4)} + [S_{11}]_{mm\,(5)} & [S_{12}]_{mm\,(5)} \\
& & & & [S_{21}]_{mm\,(5)} & [S_{22}]_{mm\,(5)} + [S_{11}]_{mm\,(6)} & [S_{12}]_{mm\,(6)} \\
& & & & & [S_{21}]_{mm\,(6)} & [S_{22}]_{mm\,(6)}
\end{bmatrix}$$

$$(8.13)$$

| Nodes I−1 | I | | | I+1....J−1 | J | | | J+1 |
|---|---|---|---|---|---|---|---|---|
| Terms | 1....m | n | r | | 1....m | n | r | |
| I−1 | | | | | | | | |
| I    1<br>  m<br>  n<br>  r | $[S_{11}]_{mn(i)}$ | | | | $[S_{12}]_{mn(i)}$ | | | |
| I+1 | | | | | | | | |
| J+1 | | | | | | | | |
| J    1<br>  m<br>  n<br>  r | $[S_{21}]_{mn(i)}$ | | | | $[S_{22}]_{mn(i)}$ | | | |
| J+1 | | | | | | | | |

$$\left. \right\} \quad (8.14)$$

Since the layers are always stacked one after the other in a sequential order (Fig. 8.1), it follows that the nodal surfaces are also numbered in a sequential order. Thus if node 1 of layer (i) is associated with node $I$ of the structure, then node 2 will always correspond to node $I+1$. The assembly of a typical submatrix $[S]_{mnpq\,(i)}$, in which

$$[S]_{mnpq\,(i)} = (\int [B]_{mn}^{T}\,[D]\,[B]_{pq}\,d\,(\text{vol.}))_{(i)}, \qquad (8.15)$$

should follow the form which is given in (8.16).

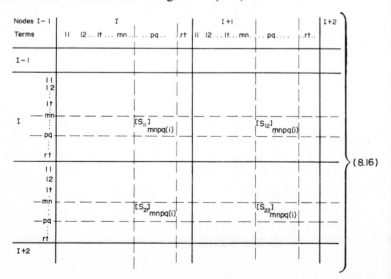

$$\left. \right\} \quad (8.16)$$

In (8.16) $r$ and $t$ are the total number of terms used for the two series respectively.

A node by node Gaussian elimination scheme, which is to be described in the next section, is used in solving the simultaneous equations. For maximum efficiency the assembly and solution processes should go hand in hand, so that the stiffness equations of a node are eliminated as soon as they are formed and no intermediate storage is required for storing the individual strip stiffness matrices. Such a solution process can be achieved simply by always having the smallest nodal number of a strip as its first nodal number and then arranging the order of the strips so that their first nodal numbers will appear in an ascending order. In this way the assemblage of the stiffness matrix of any strip will always stop automatically whenever the first nodal number of a strip is larger than the nodal number which is currently being eliminated.

## 8.5. INTRODUCTION OF PRESCRIBED DISPLACEMENTS

Unlike the finite element method, the stiffness matrix of a structure computed by the finite strip method is in general non-singular and can be solved immediately without the introduction of any boundary conditions. However, in case boundary conditions actually exist along the longitudinal edges (such as due to symmetry or support conditions), the stiffness matrix can be modified in the following manner.

Assume that we have now a set of simultaneous algebraic equations

$$
\begin{bmatrix}
k_{11} & k_{12} & k_{13} & \dots & k_{1n} \\
k_{21} & k_{22} & k_{23} & \dots & k_{2n} \\
k_{31} & k_{32} & k_{33} & \dots & k_{3n} \\
\vdots & \vdots & \vdots & & \vdots \\
k_{n1} & k_{n2} & k_{n3} & \dots & k_{nn}
\end{bmatrix}
\begin{Bmatrix}
d_1 \\ d_2 \\ d_3 \\ \vdots \\ d_n
\end{Bmatrix}
=
\begin{Bmatrix}
p_1 \\ p_2 \\ p_3 \\ \vdots \\ p_n
\end{Bmatrix}
\tag{8.17}
$$

in which one of the variables, say $d_3$, has a prescribed value of $\alpha$. The modification procedure is extremely simple and involves only two operations. Firstly, the diagonal term corresponding to $d_3$ is multiplied by a very large number which usually lies between $10^6$–$10^{10}$. Secondly, the right-hand side term $p_3$ is replaced by the product of the new diagonal coefficient and the prescribed value $\alpha$. The matrix retains its original size and arrangement, and no re-indexing, which can be quite tricky at times, is necessary.

The third equation now reads

$$k_{31} d_1 + k_{32} d_2 + k_{33} x \, 10^{10} d_3 + \ldots k_{3n} d_n = k_{33} \times 10^{10} \times \alpha. \quad (8.18)$$

Dividing throughout by $k_{33} \times 10^{10}$, we have

$$\frac{k_{31}}{k_{33} \times 10^{10}} d_1 + \frac{k_{32}}{k_{33} \times 10^{10}} d_2 + d_3 + \ldots \frac{k_{3n}}{k_{33} \times 10^{10}} d_n = \alpha. \quad (8.19)$$

Since the coefficients $k_{ij}$ usually do not differ very much in magnitude from one another, it is possible to conclude that in (8.19) all the off-diagonal coefficients are nearly equal to zero, and the equation can be rewritten as

$$d_3 = \alpha. \quad (8.20)$$

### 8.6. SOLUTION OF SIMULTANEOUS EQUATIONS

Various techniques for the solution of simultaneous equations are available but only a node by node elimination scheme will be described here. An examination of a typical stiffness matrix such as the one given in (8.13) will reveal that the matrix coefficients always cluster around the diagonal, and that beyond a certain distance from the diagonal only zero terms exist. Such a matrix is called a band matrix, and it is of the

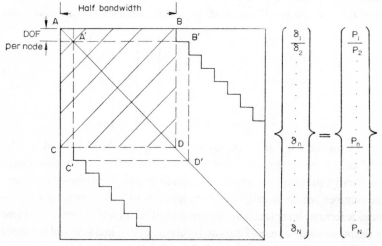

FIG. 8.2. Node by node elimination scheme. — —

utmost importance to utilize such banded property to reduce the amount of computation and also core storage requirement. The other very important property of the stiffness matrix is symmetry, i.e. $[K_{ij}] = [K_{ji}]^T$.

In the stiffness matrix shown in Fig. 8.2, each stepped area represents the equations associated with one node. The first set of nodal equations can be divided into a diagonal matrix $[K_{11}]$ and an off-diagonal submatrix $[K_{1T}] = [[K_{12}]\ [K_{13}]\ \ldots\ [K_{1n}]]$. If the nodal displacement parameters $\{\delta_1\}$ are to be eliminated, then the area $ABCD$ within the stiffness matrix will be affected and we would have the modified matrices

$$[K_{TT}]^* = [K_{TT}] - [K_{1T}]^T\ [K_{11}]^{-1}\ [K_{1T}], \qquad (8.21)$$

$$\{P_T\}^* = \{P_T\} - [K_{1T}]^T\ [K_{11}]^{-1}\ \{P_1\}, \qquad (8.22)$$

where

$$[K_{TT}] = \begin{bmatrix} [K_{22}] & \ldots & [K_{2n}] \\ \vdots & & \vdots \\ [K_{n2}] & \ldots & [K_{nn}] \end{bmatrix} \qquad (8.23)$$

and

$$\{P_T\} = \begin{Bmatrix} P_2 \\ \vdots \\ P_n \end{Bmatrix}. \qquad (8.24)$$

After the first set of nodal equations have been eliminated, the second set of equations are completed through assembly and can then be operated upon. This time the area $A'B'C'D'$ is affected, and the operations given by (8.21)–(8.24) repeated. After repeating the above-mentioned process a number of times we reach the last set of nodal equations which is of the form

$$[K_{NN}]^*\ \{\delta_N\} = \{P_N\}^*. \qquad (8.25)$$

All the modified nodal equations are stored for back-substitution purposes (Fig. 8.3).

In actual computer operations, the working store required for the elimination process is an array of the same size as $ABCD$ in Fig. 8.2. After the elimination of $[K_{11}]$ the whole $[K_{TT}]^*$ is now shifted so that $[K_{22}]^*$ now occupies the top left corner of the array. After the assembly process for node 2 has been carried out to fill up the array, the next elimination will be allowed to take place.

The last set of nodal displacement parameters $\{\delta_N\}$ can be computed from (8.25),

$$\{\delta_N\} = [K_{NN}]^{*-1} \{P_N\}^*. \tag{8.26}$$

With $\{\delta_N\}$ computed it is possible to solve the modified equations shown in Fig. 8.3 one by one in ascending order in the back-substitution process.

The modified stiffness matrix for the folded plate example is listed in (8.27). In order to save space the equations are shown in a rectangle with each diagonal submatrix starting from the same vertical position.

| $[K_{ii}]^*$ | | | | $[K_{iT}]^*$ | | | | $\{P_{ij}\}^*$ |
|---|---|---|---|---|---|---|---|---|
| $0.1713 \times 10^{12}$ | $-0.0967$ | $0.3042$ | $-0.0019$ | $-1.6919$ | $0.0967$ | $-0.3004$ | $-0.0019$ | $0$ |
| $-0.0967$ | $0.9057$ | $-0.0172$ | $0$ | $-0.0967$ | $-0.8617$ | $-0.0172$ | $0$ | $0$ |
| $0.3042$ | $-0.0172$ | $0.0586$ | $0.0108$ | $-0.3004$ | $0.0172$ | $-0.0579$ | $0.0108$ | $0.8899 \times 10^4$ |
| $-0.0019$ | $0$ | $0.0108$ | $3664 \times 10^{10}$ | $0.0019$ | $0$ | $-0.0108$ | $0.0182$ | $0.7290 \times 10^4$ |
| $1.8483$ | $-0.0876$ | $0.3044$ | $0.0561$ | $-1.6919$ | $0.0967$ | $-0.3004$ | $-0.0019$ | $0.4616 \times 10^5$ |
| $-0.0876$ | $0.9915$ | $-0.0155$ | $-0.0002$ | $-0.0967$ | $-0.8617$ | $-0.0172$ | $0$ | $-0.1276 \times 10^3$ |
| $0.3044$ | $-0.0155$ | $0.0586$ | $0.0108$ | $-0.3004$ | $0.0172$ | $-0.0579$ | $0.0108$ | $0.2669 \times 10^5$ |
| $0.0561$ | $-0.0002$ | $0.0108$ | $0.0713$ | $0.0019$ | $0$ | $-0.0108$ | $0.0182$ | $-0.1652 \times 10^4$ |
| $1.3522$ | $-0.0519$ | $0.7629$ | $0.0099$ | $-1.3077$ | $0.0850$ | $-0.7516$ | $-0.0055$ | $0.9743 \times 10^5$ |
| $-0.0519$ | $1.0602$ | $-0.0431$ | $-0.0002$ | $-0.0850$ | $-0.8596$ | $-0.0490$ | $0$ | $-0.1444 \times 10^4$ |
| $0.7629$ | $-0.0431$ | $0.4437$ | $0.0095$ | $-0.7516$ | $0.0490$ | $-0.4378$ | $0.0095$ | $0.4443 \times 10^5$ |
| $0.0099$ | $-0.0002$ | $0.0095$ | $0.0548$ | $0.0055$ | $0$ | $-0.0095$ | $0.0182$ | $-0.2428 \times 10^5$ |

$$(8.27)$$

| | | | | | | | | |
|---|---|---|---|---|---|---|---|---|
| 1.3408 | −0.0402 | 0.7677 | 0.0043 | −1.3077 | 0.0850 | −0.7516 | −0.0055 | $0.9003 \times 10^5$ |
| −0.0402 | 1.1011 | −0.0272 | 0.0092 | −0.0850 | −0.8617 | −0.0490 | 0 | $−0.1137 \times 10^5$ |
| 0.7677 | −0.0272 | 0.4464 | 0.0111 | −0.7516 | 0.0490 | −0.4378 | 0.0095 | $0.6043 \times 10^5$ |
| 0.0043 | 0.0092 | 0.0111 | 0.0571 | 0.0055 | 0 | −0.0095 | 0.0182 | $0.1640 \times 10^5$ |
| 0.1909 | 0.0663 | 0.0151 | −0.2377 | −0.1626 | 0 | 0 | −0.2432 | $0.8952 \times 10^5$ |
| 0.0663 | 3.1787 | −0.1595 | 0.0051 | 0 | −2.8990 | 0 | 0 | $−0.9836 \times 10^4$ |
| 0.0151 | −0.1595 | 5.8592 | 0.0016 | 0 | 0.1963 | −0.1963 | 0 | $0.7352 \times 10^5$ |
| −0.2377 | 0.0051 | 0.0016 | 0.5000 | 0.2432 | 0 | 5.8245 | 0.2429 | $−0.2256 \times 10^5$ |
| 0.0155 | 0.0362 | 0.0134 | 0.0465 | 0 | 0 | 0 | 0 | $0.5942 \times 10^5$ |
| 0.0362 | 0.2563 | 0.0487 | 0.1023 | 0 | 0 | 0 | 0 | $−0.7387 \times 10^5$ |
| 0.0134 | 0.0487 | 0.0140 | 0.0382 | 0 | 0 | 0 | 0 | $0.5310 \times 10^5$ |
| 0.0465 | 0.1023 | 0.0382 | 0.1552 | 0 | 0 | 0 | 0 | $0.1427 \times 10^6$ |

The nodal displacement parameters of the folded plate example are obtained from (8.27) through back-substitution.

$$\{\delta\} = [u_1 v_1 w_1 \theta_1 \quad u_2 v_2 w_2 \theta_2 \cdots u_6 v_6 w_6 \theta_6]^T$$

$$= 10^7 \times [-0.7056 \times 10^{-13} \quad 0.0582 \quad -0.6546 \quad 1658 \times 10^{-11} \quad -0.1652 \quad 0.0881 \quad 0.2354 \quad 0.2526$$
$$-0.4032 \quad 0.1284 \quad 1.5384 \quad 0.3288 \quad -1.2855 \quad 0.0583 \quad 3.0597 \quad 2.7771$$
$$-1.6546 \quad -0.0052 \quad 3.6961 \quad 0.0575 \quad -1.8251 \quad -0.4968 \quad 3.6960 \quad 0.0565]^T \qquad (8.28)$$

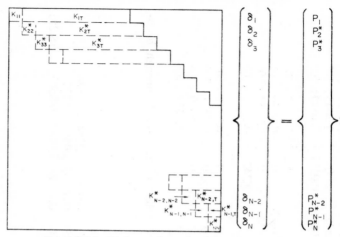

FIG. 8.3. Modified equations after forward elimination.

In order to obtain the displacement at any point along a nodal line we should multiply the relevant displacement parameter with the value of the eigenfunction (Fourier series for folded plate example) at that point.

## 8.7. CALCULATION OF INTERNAL FORCES

The stress matrices (1.21) are normally computed at the same time as the stiffness matrices and stored on disc or tape. With the displacement parameters now completely solved, such stress matrices can be retrieved from the intermediate storage for the calculation of internal forces.

For design purposes the internal forces of a strip are usually computed in its individual coordinate system. Consequently the relevant displacement parameters of the two sides of a strip should first of all be transformed to the individual coordinate system through (4.7).

The internal forces at mid-span of all the folded plate strips are listed in Table 8.1. The transverse stress $\sigma_x$ is assumed to act at the centre point of each strip section because of its stepwise variation. The shear stress $\tau_{xy}$ and torsional moment $M_{xy}$ are not listed because they are zero at the mid-span section.

TABLE 8.1. MID-SPAN STRESSES AND MOMENTS OF FOLDED PLATE

| Strip number | Transverse stress $\sigma_x$ | Longitudinal stress $\sigma_y$ | | Transverse moment $M_x$ | | Longitudinal moment $M_y$ | |
|---|---|---|---|---|---|---|---|
| | | Node 1 | Node 2 | Node 1 | Node 2 | Node 1 | Node 2 |
| 1 | −1.3368 | −0.2613 | −0.3954 | −1.5198 | 0.2020 | −0.1690 | 0.0684 |
| 2 | −1.1947 | −0.3954 | −0.5763 | 0.1988 | −0.0596 | 0.0684 | 0.4158 |
| 3 | −0.8500 | −0.5763 | −0.2616 | −0.0623 | 0.0892 | 0.4023 | 0.8634 |
| 4 | −0.3571 | −0.2616 | 0.0231 | 0.0888 | 0.0254 | 0.8636 | 1.0565 |
| 5 | −0.3903 | 0.0231 | 2.2300 | 0.0074 | 0.0006 | 3.4710 | 3.8292 |
| Multiplier | $10^4$ N/m$^2$ | $10^5$ N/m$^2$ | | $10^3$ N-m/m | | $10^2$ N-m/m | |

## 8.8. EIGENVALUE SOLUTION

In Section 5.1 it was mentioned that the free vibration of any structure can be described by the following eigenvalue equations:

$$[K]^{-1} [M] \{\delta\} = \lambda\{\delta\}. \tag{8.29}$$

However, although both $[K]$ and $[M]$ are symmetrical, the matrix product $[K]^{-1} [M]$ is in general not symmetrical, and the following transformation should be adopted.

Let the stiffness matrix $[K]$ be factorized into upper and lower triangular matrices such that

$$[K] = [L] [L]^T. \tag{8.30}$$

Inverting, we have

$$[K]^{-1} = [L]^{T-1} [L]^{-1} \tag{8.31}$$

or

$$[L]^{T-1} [L]^{-1} [M] \{\delta\} = \lambda \{\delta\}. \tag{8.32}$$

If (8.32) is now premultiplied by $[L]^T$, and taking into account that $[L]^{T-1} [L]^T = [I]$, then

$$[L]^{-1} [M] \{\delta\} = \lambda [L]^T \{\delta\}. \tag{8.33}$$

Finally, assuming that

$$\{\delta\} = [L]^{-1T} \{\delta^*\}, \tag{8.34}$$

we arrive at

$$[L]^{-1} [M] [L]^{-1T} \{\delta^*\} = \lambda [L]^T [L]^{-1T} \{\delta^*\} \tag{8.35}$$

or

$$[H] \{\delta^*\} = \lambda \{\delta^*\} \tag{8.36}$$

in which $[H]$ is now symmetrical, and therefore (8.36) can be solved economically. The actual eigenvector $\{\delta\}$ can be computed from (8.34).

## 8.9. FOLDED PLATE COMPUTER PROGRAM

The example program presented herein can be used for the analysis of a single-span folded plate with orthotropic properties. The whole program is written in Fortran IV, and its flow chart is given below.

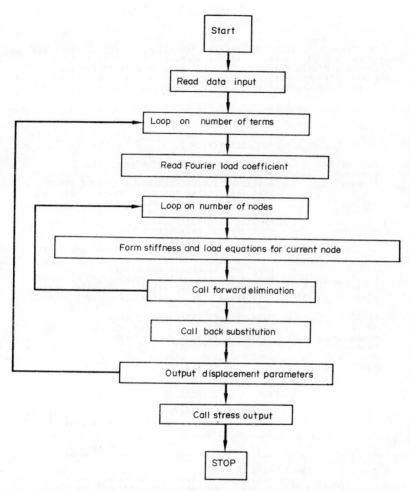

The theoretical basis of the various subroutines have been discussed in the previous sections of this chapter, and it should be possible to follow the flow of the program with the help of the many comment cards. Note that the last three subroutines INIT, STORE, and RDBK are put in primarily to save transfer time due to reading and writing of small blocks of intermediate results. For readers whose primary concern is in minimizing core storage, the three subroutines can be deleted and their functions replaced by the simple read and write statements which can be found inside the subroutines.

```
      MASTER
      DIMENSION X(50),Z(50),T(50),FORC(50),NX(20),NY(20),NL(20),FP(20)
     1,QS(4)  ,QN(2),S(4,4),C(4,4),D(4,4),E(4,4),F(4,4),QP(4)
      COMMONP(30),ST(30,30),DIS(200),R(4,4),Q(30,4),NF(8),NB(8,4)
      COMMONNTERM,NELEM,NP,NMOM,NDF,NDF1,NBAND,NSIZ,II,NI,BNII,BN1,BN2,B
      COMMON A,BB,EX,EY,PX,PY,G,DX,DY,D1,DXY,NOD(50,2),OM(600,3),DN(9)
C
C     ANALYSIS OF FOLDED PLATES AND BRIDGES
C
C
C
C     VARIABLE                DEFINITION
C
C     X(50),Z(50)             X AND Z COORDINATES OF NODAL POINTS
C     T(50)                   THICKNESS OF STRIP
C     FORC(50)                DISTRIBUTED VERTICAL LOAD ACTING ON STRIP
C     QN(1)                   LOAD COEFFICIENT FOR SINE SERIES VARIATION
C     QN(2)                   LOAD COEFFICIENT FOR COSINE SERIES VARIATION
C     ST(30,30)               FORWARD ELIMINATION WORKING AREA FOR STIFFNESS
C     P(30)                   FORWARD ELIMINATION WORKING AREA FOR LOAD
C     DIS(200)                DISPLACEMENT PARAMETER ARRAY
C     NF(8)                   RESTRAINED BOUNDARY NODE NUMBERS
C     NB(8,4)                 BOUNDARY CONDITION TYPE FOR U,V,W AND THETA
C                             (0 = FIXED, 1 = FREE)
C     NOD(50,2)               NODE NUMBERS OF ALL THE STRIPS
C     NX(20)                  NODE NUMBERS AT WHICH CONCENTRATED LOADS ACT
C     NY(20)                  CORRESPONDING Y-POSITION OF CONCENTRATED LOAD
C     NL(20)                  LOAD DEFINITION (1=FOR X-LOAD,Z-LOAD,M-LOAD,
C                             2=Y-LOAD)
C     FP(20)                  MAGNITUDE OF CONCENTRATED LOAD
C     OM(600,3)               STRESS ARRAY
C     NTERM                   NUMBER OF TERMS
C     NELEM                   NUMBER OF ELEMENTS
C     NP                      NUMBER OF NODAL LINES
C     NBOUN                   NUMBER OF RESTRAINED BOUNDARY POINTS
C     NMOM                    NUMBER OF POINT ALONG A NODAL LINE FOR OUTPUT-
C                             ING STRESSES AND MOMENTS
C     NDF                     NUMBER OF DEGREES OF FREEDOM PER NODAL LINE
C     NBAND                   MAXIMUM HALF BANDWIDTH
C     NI                      LOAD TYPE ( 1=NON-SYMMETRICAL , 2=SYMMETRICAL)
C     NCON                    NUMBER OF NODAL LINES WITH CONCENTRATED LOADS
C                             OR LINE LOADS
C
      REWIND 2
      WRITE (6,4)
    4 FORMAT( '   INPUT  DATA   ')
      READ (5,11) NTERM,NELEM,NP,NBOUN,NMOM,NDF,NBAND,NI,NCON
      WRITE(6,11) NTERM,NELEM,NP,NBOUN,NMOM,NDF,NBAND,NI,NCON
      READ  (5,35) A, (DN(I),I=1,NMOM)
      WRITE (6,35) A, (DN(I),I=1,NMOM)
      K = 4*NMOM*NELEM
      NDF1=NDF + 1
      NSIZ= NBAND - NDF
      NEQ = NDF*NP
      NA  = 2*NDF
      NCOLN = 1
      DO 123 J=1,3
      DO 123 I=1,K
  123 OM(I,J) =0
C
```

```
C       READ   X AND Z COORDINATES OF ALL POINTS
C
        DO 2  I=1,NP
        READ (5,35) X(I),Z(I)
     2  WRITE(6,35) X(I),Z(I)
C
C       READ   NUMBER NODAL NUMBERS THICKNESS VERTICAL DISTRIBUTED LOAD
C
        DO 109 I=1,NELEM
        READ (5,45) NUM,(NOD(I,J),J=1,2),T(I),FORC(I)
   109  WRITE(6,45) NUM,(NOD(I,J),J=1,2),T(I),FORC(I)
C
C       READ   BOUNDARY CONDITIONS   0=FIXED   1=FREE
C
        READ (5,19)   (NF(I),(NB(I,J),J=1,4),I=1,NBOUN)
        WRITE(6,19)   (NF(I),(NB(I,J),J=1,4),I=1,NBOUN)
    19  FORMAT(5I4)
C
C       READ   CONCENTRATED OR LINE LOAD DATA
C
        DO 222 I=1,NCON
        READ (5,45)   NX(I),NY(I),NL(I),FP(I)
   222  WRITE(6,45)   NX(I),NY(I),NL(I),FP(I)
    11  FORMAT(9I4)
    35  FORMAT(7F10.4)
    45  FORMAT(3I4,2F16.8)
C
C
C       READ   YOUNG'S MODULUS E1 E2   POISSON'S RATIO PX PY SHEAR MODULUS G
C
        READ (5,35)E1,E2,PX,PY,G
        WRITE(6,35) E1,E2,PX,PY,G
        EX = E1/(1.-PX*PY)
        EY= E2/(1.-PX*PY)
        DO  600  II=1,NTERM ,NI
        CALL INIT(NBAND,NCOLN,NDF)
        NE =0
        REWIND 4
        DO 12 I=1,NEQ
    12  DIS(I  ) = 0.
        DO  8 I=1,NBAND
        P(I)=0.
        DO 8 J=1,NBAND
     8  ST(I,J) =0.
C
C       READ   FOURIER LOAD COEFFICIENTS
C
        READ (5,35) QN(1), QN(2)
        WRITE(6,35) QN(1), QN(2)
        BNII = 3.14159*II
        CO =  BNII*BNII
        BN1 = BNII/A
        BN2 = BN1*BN1
        DO  70  LK =1,NP
        IF (LK .GT. 1) GO TO 9
     1  NE = NE + 1
        IF ( NE - NELEM) 9,9,92
     9  IF ( NOD(NE,1) - LK) 92,3,92
     3  N1 = NOD(NE,1)
        N2 = NOD(NE,2)
        XP = X(N2) - X(N1)
```

```
      ZP = Z(N2) - Z(N1)
      H = T(NE)
      B  =    SQRT(XP*XP + ZP*ZP)
      BB = B*B
      DX=E1*H**3/(12.*(1.-PX*PY))
      DY=E2*DX/E1
      DXY=G*DX/E1
      D1=PX*E2*DX/E1
      HXY=D1+4*DXY
      CALL       TRAN(  XP,ZP)
      WRITE ( 4 )((R(I,J),I=1,4),J=1,4)
      CALL       MOMP(                   D,E,F)
      CALL       MOMS(D,E,F)
      CALL REACTSX(D,E,F)
      CALL REACTSY(D,E,F)
      CALL FEMS(CO,A,B,BB,DX,DY,D1,DXY,D,E,F)
      CALL LOADS(QN,QS,NE,FORC,B,BB,R)
      CALL       FEMP(                 E,H,CO)
      CALL LOADP(QN,QP,NE,FORC,B,R)
      DO 80 LL=1,2
      J1 = 2*LL -1
      J2 = J1 + 1
      J = NDF*(NOD(NE,LL) - LK)
      S(1,1)= QP(J1)*A
      S(2,1)=QP(J2)*A
      S(3,1)= QS(J1)*A
      S(4,1)=QS(J2)*A
      CALL MBTTM(R,S,C,4,4,1)
      DO 15 NJ=1,NDF
      JN = J + NJ
  15  P(JN)   =  P(JN) + C(NJ,1)
      DO 80 KK=1,2
      DO 81 K=1,NDF
      DO 81 L=1,NDF
  81  D(K,L) =0.
      I1 = 2*KK-1
      I2 = I1 + 1
      I = NDF*(NOD(NE,KK) - LK)
      D(1,1) =   E(I1,J1)
      D(2,1) =   E(I2,J1)
      D(1,2) =   E(I1,J2)
      D(2,2) = E(I2,J2)
      D(3,3) = F(I1,J1)
      D(4,3) = F(I2,J1)
      D(3,4) = F(I1,J2)
      D(4,4) = F(I2,J2)
      CALL MBTM(D,R,C,4,4,4)
      CALL MBTTM(R,C,D,4,4,4)
      DO 5 NJ = 1,NDF
      JN =  J + NJ
      DO 5  MI = 1,NDF
      IM = MI + I
   5  ST(IM,JN) =  ST(IM,JN) + D(MI,NJ)
  80  CONTINUE
      GO TO 1
  92  DO  67 I =1,NCON
      IF (LK - NX(I)) 67,54,67
  54  J = NL(I)
      IF (NY(I)) 75,75,76
  75  IF ( J .EQ. 2) P(J) = P(J) + FP(I)*QN(2)*A
      IF ( J .NE. 2) P(J) = P(J) + FP(I)*QN(1)*A
      GO TO 67
  76  P(J) =           P(J)    +FP(I)*QN(1)
```

```
 67 CONTINUE
    CALL BOUN(LK,NBOUN)
    CALL SOLVE
 70 CONTINUE
    REWIND 4
    CALL BSUB
    CALL MOM(D,E,F,C,S)
600 CONTINUE
    K = 4*NMOM
    DO 133 LL = 1,NELEM
    N1 = (LL - 1)*K + 1
    N2 = LL*K
    WRITE(6,27)
 27 FORMAT(27H ELEMENT NUMBER NODE1 NODE2)
    WRITE(6,28)
 28 FORMAT( '   ZIGMA-X , ZIGMA-Y , ZIGMA-XY   ' )
    WRITE (6,29)
 29 FORMAT( '   MOMENT-X , MOMENT-Y , MOMENT-XY ' )
133 WRITE (6,17) LL,NOD(LL,1),NOD(LL,2),((OM(I,J),J=1,3),I=N1,N2)
 17 FORMAT(3I4/(6E13.5))
    STOP
    END



    SUBROUTINE FEMP(E,H,CO)
    DIMENSION  E(4,4)
    COMMONP(30),ST(30,30),DIS(200),R(4,4),Q(30,4),NF(8),NB(8,4)
    COMMONNTERM,NELEM,NP,NMOM,NDF,NDF1,NBAND,NSIZ,II,NI,BNII,BN1,BN2,B
    COMMON A,BB,EX,EY,PX,PY,G,DX,DY,D1,DXY,NOD(50,2),OM(600,3),DN(9)
C
C   IN-PLANE STIFFNESS MATRIX E(4,4) OF STRIP
C
    E(1,1) = EX*.5*A/B  +B*BN2*A*G/6.
    E(2,1) = .25*BN1 *A*PX*EY - .25*BN1 *A*G
    E(3,1) = -.5*A*EX/B + B*A*BN2*G/12.
    E(4,1) = .25*A*BN1 *PX*EY + .25*A*BN1 *G
    E(2,2) =A*B*BN2*EY/6. + .5*A*G/B
    E(3,2) = -.25*A*BN1 *PX*EY - .25*A*BN1 *G
    E(4,2) =A*B*BN2*EY/12. - .5*A*G/B
    E(3,3) =  .5*A*EX/B +A*B*BN2*G/6.
    E(4,3) = -.25 *A*BN1 *PX*EY + .25*A*BN1 *G
    E(4,4) =A*B*EY*BN2/6.+ .5*A*G/B
    DO  20  I=1,4
    DO  20  J=1,I
    E(I,J)=E(I,J)*H
 20 E(J,I) = E(I,J)
    RETURN
    END



    SUBROUTINE FEMS(CO,A,B,BB,DX,DY,D1,DXY,D,E,F)
    DIMENSION D(4,4),E(4,4),C(4,4),F(4,4)
C
C   BENDING STIFFNESS MATRIX F(4,4) OF STRIP
C
    AA = A*A
    BBBB= BB*BB
    COCO = CO*CO
    BBB = B*BB
```

```
      D(1,1) = COCO*DY*B*.5/AA
      D(2,1) =    COCO*BB*DY*.25/AA
      D(3,1) = -CO*B*D1   +   .166667*COCO*BBB*DY/AA
      D(4,1) = -1.5*CO*BB*D1 +   .125*COCO*BBBB*DY/AA
      D(2,2) =   .166667*COCO*BBB*DY/AA   +   2.*CO*B*DXY
      D(3,2) =-CO*BB*D1*.5+.125*COCO*BBBB*DY/AA + 2.*CO*BB*DXY
      D(4,2) = -CO*BBB*D1 +  .1*COCO*BBBB*B*DY/AA + 2.*CO*BBB*DXY
      D(3,3) =  2.*AA*B*DX -.6666667*CO*BBB*D1   +
     1 .1*COCO*BBBB*B   *DY/AA  + 2.66667*CO*BBB*DXY
      D(4,3) =  3.*AA*BB*DX   -CO*BBBB*D1 + .0833333*COCO*BBBB*BB*DY/AA
     1 +  3.*CO*BBBB*DXY
      D(4,4) =  6.*BBB*DX*AA   - 1.2*CO*BBBB*B*D1 + .0714286*COCO*BBB
     1*BBBB*DY/AA+ 3.6*CO*BB*BBB*DXY
      DO 5 I=1,4
      DO 5 J=1,I
      D(I,J) = D(I,J)/A
    5 D(J,I) = D(I,J)
      CALL EM(B,BB,BBB,E)
      CALL MBTM(D,E,C,4,4,4)
      CALL MBTTM(E,C,F,4,4,4)
      RETURN
      END


      SUBROUTINE MOMP( D,E,F)
      DIMENSION D(4,4),E(4,4),F(4,4)
      COMMONP(30),ST(30,30),DIS(200),R(4,4),Q(30,4),NF(8),NB(8,4)
      COMMONNTERM,NELEM,NP,NMOM,NDF,NDF1,NBAND,NSIZ,II,NI,BNII,BN1,BN2,B
      COMMON A,BB,EX,EY,PX,PY,G,DX,DY,D1,DXY,NOD(50,2),OM(600,3),DN(9)
C
C     IN-PLANE STRESS MATRIX D(4,4) OF STRIP
C
      DO 83 INDEX=1,2
      IF (INDEX-1) 91,91,92
   91 X = 0.
      GO TO 93
   92 X = B
   93 DO 82 M = 1,  NMOM
      Z = DN(M)/A
      H1 = SIN(BNII*Z)
      H2 = COS(BNII*Z)
      D(1,1) = -H1*EX/B
      D(2,1) = -H1*PX*EY/B
      D(3,1) = (1.-X/B)*BN1*H2*G
      D(1,2) = -(1.-X/B)*BN1*H1*PX*EY
      D(2,2) = -(1.-X/B)*BN1*H1*EY
      D(3,2) = -H2*G/B
      D(1,3) = H1*EX/B
      D(2,3) = H1*PX*EY/B
      D(3,3) = X*BN1*H2*G/B
      D(1,4) = -X*BN1*H1*PX*EY/B
      D(2,4) = -X*BN1*H1*EY/B
      D(3,4) = H2*G/B
      WRITE (4   )  ((D(I,J),I=1,3),J=1,4)
   82 CONTINUE
   83 CONTINUE
      RETURN
      END
```

```
      SUBROUTINE MOMS(D,E,F)
      DIMENSION D(4,4),E(4,4),F(4,4)
      COMMONP(30),ST(30,30),DIS(200),R(4,4),Q(30,4),NF(8),NB(8,4)
      COMMONNTERM,NELEM,NP,NMOM,NDF,NDF1,NBAND,NSIZ,II,NI,BNII,BN1,BN2,B
      COMMON A,BB,EX,EY,PX,PY,G,DX,DY,D1,DXY,NOD(50,2),OM(600,3),DN(9)
C
C     MOMENT MATRIX F(4,4) OF STRIP
C
      CO = BN2
      DO83 INDEX=1,2
      IF (INDEX-1) 91,91,92
   91 X = 0.
      GO TO 93
   92 X = B
   93 DO 82 M= 1,   NMOM
      Z = DN(M)/A
      H1  = SIN(BNII*Z)
      H2  =COS(BNII*Z)
      H3  =-H1
      D(1,1) = -CO*H3*D1
      D(2,1) = -CO*H3*DY
      D(3,1) = 0
      D(1,2) = -CO*H3*D1*X
      D(2,2) = -CO*X*H3*DY
      D(3,2) = 2.*BN1 *H2*DXY
      D(1,3) = -2.* H1*DX- CO*X*X*H3*D1
      D(2,3) = -2,*H1 *D1 - CO*X*X*H3*DY
      D(3,3) = 4.*BN1 *H2*DXY*X
      D(1,4) = -6.*X*H1*DX - CO*X*X*H3*D1*X
      D(2,4) = -6.*X*H1 *D1 - CO*X*X*X*H3*DY
      D(3,4) =  6.*BN1  *X*X*H2*DXY
      BBB = BB*B
      CALL EM(B,BB,BBB,E)
      CALL MBTM(D,E,F,3,4,4)
   82 WRITE (4  )  ((F(I,J),I=1,3),J=1,4)
   83 CONTINUE
      RETURN
      END




      SUBROUTINE TRAN(XP,ZP)
      COMMONP(30),ST(30,30),DIS(200),R(4,4),Q(30,4),NF(8),NB(8,4)
      COMMONNTERM,NELEM,NP,NMOM,NDF,NDF1,NBAND,NSIZ,II,NI,BNII,BN1,BN2,B
      COMMON A,BB,EX,EY,PX,PY,G,DX,DY,D1,DXY,NOD(50,2),OM(600,3),DN(9)
C
C     TRANSFORMATION MATRIX R(4,4) OF STRIP
C
      DO 1 I=1,4
      DO 1 J=1,4
    1 R(I,J) =0.
      S = ZP/B
      C = XP/B
      R(1,1) = C
      R(2,2) = 1.
      R(3,1) = -S
      R(1,3) = S
      R(3,3) = C
```

```
      R(4,4) = 1.
      RETURN
      END




      SUBROUTINE LOADP(QN,QP,LK,FORC,B,R)
      DIMENSION QN(2),QP(4),FORC(99),R(4,4)
C
C     IN-PLANE NODAL FORCES QP(4) DUE TO DISTRIBUTED LOADS
C
      FOR = FORC(LK)*R(1,3)
      QP(1) =QN(1)*FOR*B*.5
      QP(2) =0.
      QP(3) =QP(1)
      QP(4) = 0.
      RETURN
      END




      SUBROUTINE LOADS(QN,QS,LK,FORC,B,BB,R)
      DIMENSION QN(2),QS(4),FORC(99),R(4,4)
C
C     OUT-OF-PLANE NODAL FORCES QS(4) DUR TO DISTRIBUTED LOADS
C
      FOR= FORC(LK)*R(1,1)
      DO 26 KK = 1,2
      I = 2*(KK-1)
      QS(I+1) = QN(1)*FOR*B*.5
      GO TO (24,25),KK
   24 QS(I+2) = QN(1)*BB*FOR/12.
      GO TO 26
   25 QS(I+2) =-QN(1)*BB*FOR/12.
   26 CONTINUE
      RETURN
      END




      SUBROUTINE BOUN( LK,NBOUN)
      COMMONP(30),ST(30,30),DIS(200),R(4,4),Q(30,4),NF(8),NB(8,4)
      COMMONNTERM,NELEM,NP,NMOM,NDF,NDF1,NBAND,NSIZ,II,NI,BNII,BN1,BN2,B
      COMMON A,BB,EX,EY,PX,PY,G,DX,DY,D1,DXY,NOD(50,2),OM(600,3),DN(9)
C
C     INTRODUCTION OF BOUNDARY CONDITIONS
C
      DO 230 I=1,NBOUN
      IF (NF(I) -LK) 78,79,78
   79 DO.230 J=1,NDF
      IF (NB(I,J)) 230,345,230
  345 ST(J,J) = ST(J,J)*.1E+12
   78 CONTINUE
  230 CONTINUE
      RETURN
      END
```

```
      SUBROUTINE SOLVE
      COMMONP(30),ST(30,30),DIS(200),R(4,4),Q(30,4),NF(8),NB(8,4)
      COMMONNTERM,NELEM,NP,NMOM,NDF,NDF1,NBAND,NSIZ,II,NI,BNII,BN1,BN2,B
      COMMON A,BB,EX,EY,PX,PY,G,DX,DY,D1,DXY,NOD(50,2),OM(600,3),DN(9)
C
C     FORWARD ELIMINATION
C
      CALL MATIN(ST,NDF)
      CALL STORE(ST,P,NDF)
      DO 111 J=1,NDF
      DO 111 I=1,NSIZ
      L = I + NDF
      Q(I,J) = 0.
      DO 111 K=1,NDF
  111 Q(I,J) = Q(I,J) + ST(K,L)*ST(K,J)
      DO 112 I=   NDF1,NBAND
      L = I- NDF
      DO 112 K=1,NDF
      DO 113 J=NDF1,NBAND
  113 ST(I,J) = ST(I,J) - Q(L,K)*ST(K,J)
  112 P(I   ) = P(I   ) - Q(L,K)*P(K   )
      DO 114 I=1,NSIZ
      K = I + NDF
      P(I   ) = P(K   )
      P(K   ) = 0.
      DO 114 J=1,NSIZ
      L = J+NDF
      ST(I,J) = ST(K,L)
      ST(I,L) = 0.
      ST(K,J) = 0.
  114 ST(K,L) = 0.
      RETURN
      END


      SUBROUTINE BSUB
      COMMONP(30),ST(30,30),DIS(200),R(4,4),Q(30,4),NF(8),NB(8,4)
      COMMONNTERM,NELEM,NP,NMOM,NDF,NDF1,NBAND,NSIZ,II,NI,BNII,BN1,BN2,B
      COMMON A,BB,EX,EY,PX,PY,G,DX,DY,D1,DXY,NOD(50,2),OM(600,3),DN(9)
C
C     BACK-SUBSTITUTION
C
      DO 30IP=1,NP
      M = NP     - IP
      CALL   RDBK(ST,P,NDF)
      DO 1  I=1,NDF
      DO 1 J=NDF1,NBAND
    1 P(I   ) = P(I   ) - ST(I,J) * P(J   )
      DO 2  I=1,NDF
      Q(I, 1) = 0.
      DO 2 J=1,NDF
    2 Q(I, 1) = Q(I, 1) +ST(I,J)*P(J   )
      DO 3 I=1,NDF
      P(I) = Q(I,1)
      J = NDF*M + I
    3 DIS(J) = Q(I,1)
      DO 114 I=1,NSIZ
      L = NBAND - I + 1
```

```
      K = L - NDF
114 P(L) = P(K)
 30 CONTINUE
    RETURN
    END



    SUBROUTINE MOM(D,E,F,C,S)
    DIMENSION D(4,4),E(4,4),F(4,4),C(4,4),S(4,4)
    COMMONP(30),ST(30,30),DIS(200),R(4,4),Q(30,4),NF(8),NB(8,4)
    COMMONNTERM,NELEM,NP,NMOM,NDF,NDF1,NBAND,NSIZ,II,NI,BNII,BN1,BN2,B
    COMMON A,BB,EX,EY,PX,PY,G,DX,DY,D1,DXY,NOD(50,2),OM(600,3),DN(9)
C
C
C   COMPUTATION OF STRESSES (D=FC), MOMENTS (D=FS) AT SPECIFIED POINTS
C   ALONG NODAL LINES AND DISPLACEMENTS DIS(200) AT CENTRE POINTS
C
    NUM = 0
    WRITE(6,10) II
 10 FORMAT(I4)
    DO 81 LK=1,NELEM
    READ (4  ) ((R(I,J),I=1,4),J=1,4)
    DO 24 LL=1,2
    I = NDF*(NOD(LK,LL)- 1)
    DO 23 J=1,NDF
    K = I + J
 23 D(J,1) = DIS(K)
    CALL MBTM(R,D,E,4,4,1)
    C(2*LL-1,1) = E(1,1)
    C(2*LL,1)   = E(2,1)
    S(2*LL-1,1) = E(3,1)
    S(2*LL,1)   = E(4,1)
 24 CONTINUE
    DO 25 LL=1,2
    DO 25 M=1,  NMOM
    READ (4  ) ((F(I,J),I=1,3),J=1,4)
    CALL MBTM(F,C,D,3,4,1)
    NUM = NUM + 1
    DO 3 I=1,3
  3 OM (NUM,I) = OM(NUM,I) + D(I,1)
 25     CONTINUE
    DO 26 LL=1,2
    DO 26 M=1,  NMOM
    READ (4  ) ((F(I,J),I=1,3),J=1,4)
    CALL MBTM(F,S,D,3,4,1)
    NUM = NUM + 1
    DO 4 I=1,3
  4 OM (NUM,I) = OM(NUM,I) + D(I,1)
 26     CONTINUE
 81 CONTINUE
    NP2 = NDF*NP
    DO 7 I =1,NP2,NDF
    DIS(I  ) = DIS(I  )*SIN(BNII*.5)
    DIS(I+2) = DIS(I+2)*SIN(BNII*.5)
  7 DIS(I+3) = DIS(I+3)*SIN(BNII*.5)
    WRITE (6,15)
 15 FORMAT(11H DEFLECTION,9H ROTATION)
    WRITE (6,5) (DIS(I  ),I=1,NP2)
  5 FORMAT(4E16.8)
    RETURN
    END
```

```
      SUBROUTINE EM(B,BB,BBB,E)
      DIMENSION E(4,4)
      E(1,1) = 1.
      E(2,1) = 0.
      E(3,1) = -3./BB
      E(4,1) = 2./BBB
      E(1,2) = 0.
      E(2,2) = 1.
      E(3,2) = -2./B
      E(4,2) = 1./BB
      E(1,3) = 0.
      E(2,3) = 0.
      E(3,3) = 3./BB
      E(4,3) = -2./BBB
      E(1,4) = 0.
      E(2,4) = 0.
      E(3,4) =-1./B
      E(4,4) = 1./BB
      RETURN
      END
```

```
      SUBROUTINE MATIN(ST,N)
      DIMENSION ST(30,30)
C
C     MATRIX INVERSION
C
      DO 19 I=1,N
      Z=ST(I,I)
      ST(I,I)=1.
      DO 60 J=1,N
   60 ST(I,J)=ST(I,J)/Z
      DO 19 K=1,N
      IF(K-I)3,19,3
    3 Z=ST(K,I)
      ST(K,I)=0
      DO 4 J=1,N
    4 ST(K,J)=ST(K,J)-Z*ST(I,J)
   19 CONTINUE
      RETURN
      END
```

```
      SUBROUTINE MBTM (D,B,DB,L,M,N)
      DIMENSION D(4,4),B(4,4),DB(4,4)
C
C     MATRIX MULTIPLICATION   DB(L,N) = D(L,M) X B(M,N)
C
      DO 110 J=1,N
      DO 110 I=1,L
      DB(I,J)=0.
      DO 110 K=1,M
  110 DB(I,J)=DB(I,J)+D(I,K)*B(K,J)
      RETURN
      END
```

```
      SUBROUTINE MBTTM(D,B,DB,L,M,N)
      DIMENSION D(4,4),B(4,4),DB(4,4)
C
C     MATRIX TRANSPOSE MULTIPLICATION  DB(L,N) = D(M,L) X B(M,N)
C
      DO 110 J=1,N
      DO 110 I=1,L
      DB(I,J)=0.
      DO 110 K=1,M
  110 DB(I,J)=DB(I,J)+D(K,I)*B(K,J)
      RETURN
      END
```

```
      SUBROUTINE INIT ( NBAND, NCOLN ,NDF )
C          THIS SUBROUTINE MUST BE CALLED ONCE BEFORE SUBROUTINES
C     STORE AND RDBK ARE CALLED IN ORDER TO INITIALISE THE BLOCK
C     CONTROL COUNTERS.
      COMMON/BUFDA/NBD,NCOL,IS,NA,LRECL,NREC,L,X(2000)
      NA = 2000
      IS = 1
      NBD = NBAND
      NCOL = NCOLN
      LRECL =(NBD + NCOL )*NDF
      NREC = 0
      IF ( LRECL - NA ) 1, 1, 2
    1 RETURN
    2 WRITE ( 6,4) LRECL, NA
    4 0FORMAT ('0LOGICAL RECORD LENGTH OF ',I6,'EXCEEDS BUFFER SET AT',
     1 I6 )
      STOP
      END
```

```
      SUBROUTINE STORE (ST,P,NDF)
      DIMENSION ST(30,30),P(30)
      COMMON/BUFDA/NBD,NCOL,IS,NA,LRECL,NREC,L,X(2000)
C
C     STORE FORWARD ELIMINATION RESULTS IN BUFFER AREA AND WRITE ON TAPE
C     IN A BLOCK OF 2000 WORDS TO SAVE TRANSFER TIME
C
C     TEST IF ROOM IN CURRENT BUFFER
      IF (IS + LRECL - NA ) 5, 5, 50
C     ROOM IN BUFFER
    5 DO 10 I = 1, NBD
      DO 10 J=1,NDF
      X(IS) = ST (J,I)
   10 IS = IS + 1
      DO 15 I = 1,NDF
      X (IS) = P( I)
   15 IS = IS + 1
      RETURN
C     NO ROOM LEFT IN BUFFER
```

```
50     L = IS - 1
       WRITE (2) (X(J), J = 1,L )
       IS = 1
       NREC = NREC + 1
       GO TO 5
       END


       SUBROUTINE RDBK (ST, P, NDF)
       DIMENSION ST(30,30),P(30)
       COMMON/BUFDA/NBD,NCOL,IS,NA,LRECL,NREC,L,X(2000)
C
C      READ FORWARD ELIMINATION RESULTS FOR BACK-SUBSTITION
C
C      TEST IF NEXT RECORD IN BUFFER
10     IS = IS - LRECL
       IF ( IS - 1 ) 40,12,12
C      RECORD IS IN BUFFER
12 DO 11 I = 1, NBD
       DO 11 J=1,NDF
       ST ( J, I) = X(IS)
11     IS = IS + 1
       DO 15 I = 1, NDF
       P ( I) = X(IS)
15     IS = IS + 1
       IS = IS      - LRECL
       RETURN
C      LAST BLOCK WRITTEN MUST BE READ
40     IF (NREC) 100,100,41
41     NREC = NREC - 1
       BACKSPACE   2
       READ ( 2) ( X(J), J = 1,L)
       BACKSPACE 2
       IS = L + 1
       GO TO 10
C      ILLOGICAL ERROR
100    WRITE (6,101)
101    FORMAT ('0 ATTEMPT TO READ BACK TOO MANY RECORDS.')
       STOP
       END


       INPUT   DATA
       1   5   6   1   1   4   8   1   1
       70.0000    35.0000
       0.0000     0.0000
       4.9150      .8750
       9.8300     1.7500
      14.1650     4.2500
      18.5000     6.7500
      18.5000     9.7500
       1   1   2        .25000000    80.00000000
       2   2   3        .25000000    80.00000000
       3   3   4        .25000000    80.00000000
       4   4   5        .25000000    80.00000000
       5   5   6        .50000000    75.00000000
       1   0   1   1   0
       1   1   3      0.00000000
       1.0000     1.0000   -0.0000    -0.0000     .5000
```

```
      .6366    -0.0000
    1
DEFLECTION ROTATION
  -.70558786E-06     .58219091E+06   -.65459828E+07     .16574907E-04
  -.16522500E+07     .88095188E+06    .23541902E+07     .25261783E+07
  -.40323523E+07     .12840635E+07    .15383274E+08     .32878075E+07
  -.12855114E+08     .58290072E+06    .30596795E+08     .27711330E+07
  -.16545901E+08    -.51553719E+05    .36960847E+08     .57525700E+06
  -.18250393E+08    -.49675224E+07    .36959676E+08     .56466399E+06
ELEMENT NUMBER NODE1 NODE2
ZIGMA -X , ZIGMA -Y , ZIGMA -XY
MOMENT-X , MOMENT-Y , MOMENT-XY
    1    1    2
  -.13368E+05  -.26129E+05    .55414E-02   -.13368E+05  -.39537E+05    .35545E-02
  -.15198E+04  -.16902E+02    .12851E-14    .20200E+03   .68382E+01    .19587E-03
ELEMENT NUMBER NODE1 NODE2
ZIGMA -X , ZIGMA -Y , ZIGMA -XY
MOMENT-X , MOMENT-Y , MOMENT-XY
    2    2    3
  -.11947E+05  -.39537E+05    .17421E-01   -.11947E+05  -.57629E+05    .15645E-01
   .19882E+03   .68382E+01    .19587E-03   -.59611E+03   .41574E+02    .25492E-03
ELEMENT NUMBER NODE1 NODE2
ZIGMA -X , ZIGMA -Y , ZIGMA -XY
MOMENT-X , MOMENT-Y , MOMENT-XY
    3    3    4
  -.85004E+04  -.57629E+05    .31859E-01   -.85004E+04  -.26161E+05    .30593E-01
  -.62257E+03   .40233E+02    .25492E-03    .89145E+03   .86357E+02    .21486E-03
ELEMENT NUMBER NODE1 NODE2
ZIGMA -X , ZIGMA -Y , ZIGMA -XY
MOMENT-X , MOMENT-Y , MOMENT-XY
    4    4    5
  -.35713E+04  -.26161E+05    .39436E-01   -.35713E+04   .23137E+04    .38904E-01
   .88839E+03   .86357E+02    .21486E-03    .25433E+03   .10565E+03    .44602E-04
ELEMENT NUMBER NODE1 NODE2
ZIGMA -X , ZIGMA -Y , ZIGMA -XY
MOMENT-X , MOMENT-Y , MOMENT-XY
    5    5    6
  -.39030E+03   .23137E+04    .13362E-01   -.39030E+03   .22294E+06    .13327E-01
   .74208E+02   .34715E+03    .35682E-03   -.64545E+00   .38292E+03    .35025E-03
```

# Author Index

Page numbers in italic type indicate References at end of chapters

# Subject Index

# A Selection of Important Pergamon Journals of Interest

Building and Environment

Computers and Structures

Cement and Concrete Research

Engineering Fracture Mechanics

International Journal of Engineering Science

International Journal of Non-Linear Mechanics

International Journal of Solids and Structures

The Journal of Physics and Chemistry of Solids

Journal of the Mechanics and Physics of Solids

Materials and Society

Materials Research Bulletin

Mechanics Research Communications

Solar Energy

Thermal Engineering

444